MATHEMATICS
FOR
BUSINESS ANALYSIS

Mathematics for Business Analysis

Paul Turner, PhD
and
Justine Wood, PhD

Mercury Learning and Information
Boston, Massachusetts

Publisher: David Pallai
MERCURY LEARNING AND INFORMATION
121 High Street, 3rd Floor
Boston, MA 02210
info@merclearning.com
www.merclearning.com
800-232-0223

P. Turner and J. Wood. *Mathematics for Business Analysis*.
ISBN: 978-1-68392-937-6

Library of Congress Control Number: 2023944273

232425321 Printed on acid-free paper in the United States of America.

I would like to dedicate this book to my Mum and Dad who have given me all the support I could possibly ask for throughout my academic career.

—Paul Turner

I dedicate this book to my parents, for their continuous love and support.

—Justine Wood

CONTENTS

PREFACE

In developing this book, we have drawn on our experiences of teaching mathematics to economics and business students over a long period of time. This is, more often than not, a challenging task because mathematics can be viewed as unpopular with students, who may regard it as a chore rather than a pleasure. Nevertheless, we believe that teaching mathematics as part of economics and business programs can be of immense value to the students concerned and can even be enjoyable for the staff involved. What is needed is a clear program of study and a willingness to explain the subject from basics rather than as just a set of unrelated techniques. This is what we attempt to do in this book.

Our approach is as follows: first, in Chapter 1, we develop the very basics of mathematics in terms of the nature of numbers, starting with the natural numbers and progressing to the integers, real numbers and finally, introducing more exotic concepts such as complex numbers and the hyperreals. This naturally allows us to develop the idea of sets which act as a basic organizing structure in mathematics. Chapter 2 then builds on this to develop the idea of mathematical functions as a 'mapping' from one set to another. Much of this initial material is designed to allow students to become comfortable with the language of mathematics and to enable them to express familiar concepts in a more formal manner.

The basic material of Chapters 1 and 2 is then followed by applications which make use of mathematical functions to address topics of interest for students of economics and business. In Chapter 3, we look at the solution of systems of simultaneous equations. This has obvious applications in the analysis of interactions between economic agents and the determination of market equilibrium. We consider methods for the solution of systems of equations and show how these can be applied to both linear and non-linear systems. Our initial treatment of this topic is limited to small systems containing only two or three equations, but this is later extended in Chapter 8 when we introduce the method of matrices as a way of extending our solution methods to larger systems.

Chapters 4 to 7 comprise a largely self-contained section which can be used as the basis for a course in elementary calculus. Chapter 4 introduces both the idea of the derivative of a function and covers the standard methods of differentiation. We then use this in Chapter 5, to develop methods for finding maximum and minimum points of functions. In particular, we apply these methods to standard problems in economics and business such as finding profit maximizing levels of output or cost minimizing combinations of factors of production. In Chapter 5, we limit this discussion to the case of functions with a single input variable. In Chapter 6 however, we extend this to deal with multivariable functions which allow for multiple inputs. We also introduce the idea of constraints to optimization problems which require the use of Lagrangian methods. At all stages, we develop the mathematical discussion using examples drawn from economics and business to illustrate the relevance of these methods to problems of interest for students. The calculus section is completed in Chapter 7, with an introduction to integral calculus and the process of integration. Again, we take care to develop the methods we introduce using examples of interest drawn from the relevant literature.

A novel feature of our treatment of calculus is the use of infinitesimal methods. This differs from the standard treatment in many textbooks which typically use the method of limits to develop both derivative and integral calculus. The use of infinitesimal methods requires some initial investment in technique in that it requires the use of hyperreal numbers, which we introduce in Chapter 1. These are numbers which are either infinitesimally small, that is smaller than any non-zero real number, or infinitely large, that is greater than any real number. However, we believe that this framework offers significant advantages over the conventional limits approach in terms of increased intuition and ease of development of methods for the processes of differentiation and integration.

Chapters 1 to 7 cover most of the essential material for an introductory undergraduate module in calculus for economics and business studies. Most such programs will, however, find it useful to introduce more advanced mathematical methods at a later stage. In Chapters 8 to 11, we therefore cover a number of topics which feature in the later stage of undergraduate programs and in master's programs. Chapter 8 introduces the use of matrix methods to solve systems of equations which generalizes the introductory material of Chapter 2 to permit the solution of simultaneous systems consisting of many equations. Finally, in Chapters 9, 10, and 11, we introduce the idea of differential and difference equations. These are systems of equations which allow for the analysis of dynamic systems, that is variables which change through time in response to external stimulus. As with earlier chapters, we illustrate the utility of these methods using economics or business examples at every stage.

A novel feature of our approach is the integration of numerical methods throughout the book. We do this using computer code written in the PYTHON computing language. This allows many of our examples to be illustrated numerically, which we believe helps students both understand the material more clearly and appreciate how it can be applied in practical situations. The code for our applications is provided in all cases and is available for teachers to both use and adapt as they wish. Companion files from the book are available by writing to the publisher at info@merclearning.com.

We would like to acknowledge the input of Jim Walsh, Shane Stanton, and Jennifer Blaney for help in turning the manuscript into a finished product with the usual proviso that any remaining errors are the responsibility of the authors.

Our book has been developed based on our experience in teaching mathematics to students on a wide range of different programs. It reflects what we have found to be useful and interesting for students. We hope very much that users of this book, whether teachers or students, find our approach to be of use.

Paul Turner
Justine Wood
October 2023

CHAPTER 1

SETS, NUMBERS, AND ALGEBRA

Numbers are the raw material of mathematics. In this chapter, we define the types of numbers that you will encounter as part of your studies. To do this, we make use of the concept of a *set* or collection of objects—which is fundamental in mathematics. We also discuss the rules of arithmetic and algebra, allowing us to manipulate mathematical objects consistently.

1.1 SETS AND NUMBERS

The idea of a *set* in mathematics is a very general concept that includes any collection of objects. In mathematics, we are particularly interested in sets consisting of *numbers*, where a number is a mathematical object which we use to count, label, or measure other objects. The simplest numbers are the *counting* or *whole* numbers, 1, 2, 3, etc. We can define a set as a collection of objects with a rule for determining which objects belong to the set and which do not. For example, suppose we define set A to be the set of positive whole numbers less than four. This can be written in mathematical notation as $A=\{1,2,3\}$, where the *elements* of the set are listed between curly parentheses, also known as curly brackets or braces. For small sets, we can simply list all the elements. However, this becomes cumbersome when sets become larger, and impossible when there is an infinite number of elements. A set is described as finite when the number of elements is limited and infinite when the number of elements is unlimited. The set A is finite because it contains only three elements, but it is easy to define sets which contain an infinite number of elements. For example, let B be the set of all positive whole numbers greater than 3, i.e., $B=\{4,5,6,\ldots\}$. The ellipsis, or dots, in this expression indicates that there are further elements in this set that increase according to the rule established by the elements shown. That is each new element increases by one relative to the preceding element.

A set is said to be *well-defined* if there is a clear rule for deciding whether a particular object is an element of the set. For example, in the case of A, it is clear that the number 2 is an element, but the number 5 is not. Similarly, in the case of B, it is clear by the definition that the number 2 does not belong in the set whereas the number 100 does. Defining a set in terms of a rule is often easier than simply listing its elements. Set theory allows the elements of a set to consist of any type of object, providing we can define rules for their inclusion or exclusion. For example, the set of *additive primary colors* consists of three colors, red, green, and blue, which can be mixed to produce almost any other color. We can define this as the set $C = \{\text{red, green, blue}\}$. Again, there is a clear rule for determining which colors belong in the set and which do not.

The first set of numbers of interest to us is the set of *natural numbers*. This is an infinite set which consists of the numbers we use for counting purposes. We write this set as:

$$\mathbb{N} = \{1, 2, 3, \ldots.\}. \tag{1.1}$$

Note that we can form the set of natural numbers by merging the sets A and B, which we defined earlier. This defines the *union* of the two sets and is written as $\mathbb{N} = A \cup B$. If a number x is an element of either of the sets A or B, then it is, by definition, an element of the set $\mathbb{N}$. Since the set B is an infinite set, it follows that the set $\mathbb{N}$ is also infinite. This set is sometimes referred to as the *counting numbers* since it comprises the basic numbers used to count other objects.

Set theory has an associated notation; it is important to become familiar with its conventions. We have already made use of the symbol $\cup$, which means the *union* of two sets, that is, a set that contains all the elements of two other sets. Similarly, the symbol $\cap$ is used to mean the *intersection* of two sets, that is, the elements which are present in both sets. A *Venn diagram* provides a useful way of illustrating and understanding this distinction. In Figure 1.1, we have two sets of numbers $A = \{1, 2, 3, 4\}$ and $B = \{4, 5, 6, 7\}$, which are shown as being contained with circles. The union of these sets consists of all numbers which are contained in *either* of the two sets, that is, $A \cup B = \{1, 2, 3, 4, 5, 6, 7\}$, while the intersection of the sets consists of the single number 4, which is the only number that is an element of both sets, so $A \cap B = \{4\}$.

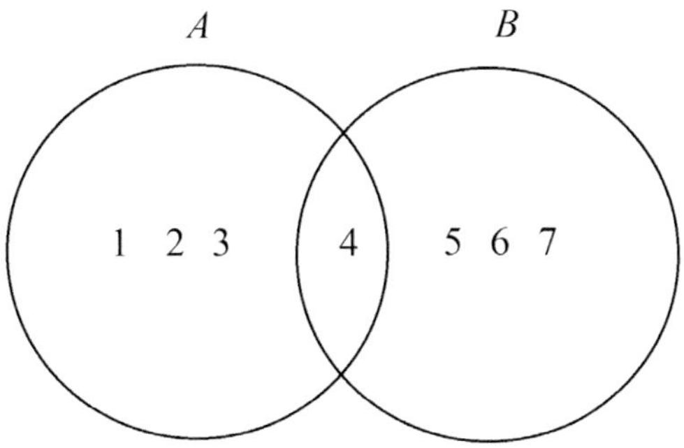

$$A \cup B = \{1,2,3,4,5,6,7\}$$

$$A \cap B = \{4\}$$

FIGURE 1.1 Venn diagram representation of sets.

In some cases, there may be no intersection between sets. For example, let $A = \{1,2\}$ and $B = \{3,4\}$. These sets have no elements in common. In situations like this, the intersection of the two sets defines the *null* or *empty* set. This is a set that contains no elements and is written as $A \cap B = \varnothing$. Another way to describe this situation is to say that sets A and B are *distinct* sets or *mutually exclusive*, in the sense that they have no common elements. Note that the empty set does not contain the number zero. If zero is a common element of two sets, then their intersection cannot be said to be empty.

Another useful item of notation is the symbol $\subseteq$ which is used to indicate that one set is a *subset* of another set. For example, if $A = \{1,2\}$ and $B = \{1,2,3,4\}$ then all the elements of A are present in B and, therefore, A is a subset of B. This is written as $A \subseteq B$ and, by definition, this makes B a *superset* of A, which we write as $B \supseteq A$. Note that this definition of a subset allows the case in which the sets are simply identical, i.e., $A = B$. If we modify the symbol to exclude the horizontal line, then a statement of the form $A \subset B$ indicates that A is a *proper subset* of B, i.e., all the elements of A are present in B, and there is at least one element of B that is *not* present in A. For example, if $A = \{1,2\}$ and $B = \{1,2,3\}$, then A is a proper subset of B because all the elements of B are also present in A, and the number 3 is present in the set B but not in set A. A line through this symbol indicates the opposite interpretation. For example, $A \not\subset B$ means that A is not a proper subset of B. This would be the case, for example, if $A = \{1,2\}$ and $B = \{2,3\}$ because the number 1 is an element of set A but not of set B.

Another symbol that you will see frequently is $\in$, which is used to indicate that an element is present in a set. That is, the statement $x \in A$ indicates that the object x is an element of the set A. For example, the number 100 is a natural number, and we can therefore write $100 \in \mathbb{N}$. On the other hand, the fraction ½ is not a natural number, and we would therefore use the symbol $\notin$ to indicate that it does not belong in this set, i.e., $1/2 \notin \mathbb{N}$. In general, $x \notin A$ can be read as "x is not an element of the set A."

At this stage, we have introduced quite a lot of new concepts and associated notation. It is, therefore, useful to consolidate this new information and provide some examples. Table 1.1 provides a summary of the set definitions we have introduced so far and gives examples of the standard notation, which should help to clarify these definitions.

TABLE 1.1 Set theory notation.

Description	Notation	Examples
Union of two sets All elements that are in set A or set B or both sets.	$A \cup B$	$A = \{1,2,3,4,5\}$ $B = \{3,4,5,6,7\}$ $A \cup B = \{1,2,3,4,5,6,7\}$
Intersection of two sets All elements that are present in both sets.	$A \cap B$	$A = \{1,2,3,4,5\}$ $B = \{3,4,5,6,7\}$ $A \cap B = \{3,4,5\}$
Subset A set A is a subset of set B if all elements in A are also present in B. (This definition allows $A = B$)	$A \subseteq B$	$A = \{1,2\}$ $B = \{1,2,3,4,5\}$ $A \subseteq B$ $A = \{1,2\}$ $B = \{1,2\}$ $A \subseteq B$ and $B \subseteq A$ In both these examples B is a superset of A
Proper subset A set A is a proper subset of set B if all elements in A are also present in B, and there is at least one element of B that is not in A. (This definition excludes $A=B$)	$A \subset B$	$A = \{1,2\}$ $B = \mathbb{N}$ $A \subset B$ In this example B is a proper superset of A
Elements of a set The element x is present in the set A.	$x \in A$	$A = \{1,2\}$ If $x = 1$ or $x = 2$ then $x \in A$. If $x = 3$ then $x \notin A$
Set difference or relative complement All the elements of the more general set that are not present in set A.	B-A	$A = \{1,2\}$ $B = \{1,2,3,4,5\}$ $B - A = \{3,4,5\}$
Null set	$\varnothing = \{\ \}$	The null set contains no elements.

A set is said to be *closed* for a mathematical operation if the application of that operation to two or more of its elements creates a third element which is also an element of the original set. For example, the set of natural numbers is closed under both addition and multiplication. This means that if x and y are natural numbers ($x, y \in \mathbb{N}$) then it is always the case that $x + y \in \mathbb{N}$ and $xy \in \mathbb{N}$. However, the set of natural number is not closed under subtraction or division. This is easily demonstrated by providing contradictory examples. For instance, $2 - 3 = -1$, which demonstrates that the set of natural numbers is not closed under subtraction because negative numbers are not contained within the set $\mathbb{N}$. Similarly, we have $2 / 4 = 1 / 2$, which is not a natural number, and therefore establishes that $\mathbb{N}$ is not closed under division.

Although the set of natural numbers is not closed under the operation of subtraction, we can define a new set that has this property. This set is referred to as the set of *integers* and is generally described using the symbol $\mathbb{Z}$. The set of *integers* includes all the natural numbers, as well as the number zero and the negative counterparts of the natural numbers. It can therefore be written as

$$\mathbb{Z} = \{...., -2, -1, 0, 1, 2, ...\} \tag{1.2}$$

Note that the set of natural numbers is a proper subset of the set of integers because every element of the set of natural numbers is also an element of the set of integers, but there are integers that are not natural numbers. In mathematical notation, this relationship is written as $\mathbb{N} \subset \mathbb{Z}$. The set of integers is closed under subtraction because if x and y are integers, then $x - y$ will also be an integer. However, the set of integers is not closed under division, as we have already demonstrated using the example. $2 / 4 = 1 / 2 \notin \mathbb{Z}$.

A useful way to think of the integers is as evenly spaced points lying along an infinitely long line, as illustrated in Figure 1.2.

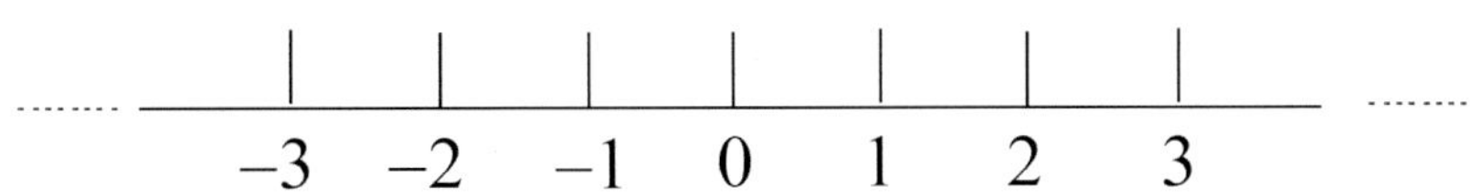

FIGURE 1.2 The number line showing integers.

This line extends infinitely in both directions from point 0, which we refer to as the *origin*. The number line is useful because it gives us a visual representation of some of the basic operations of arithmetic. We can think of the operation of addition as a rightward movement along this line. Adding the number two to the number one means starting at point 1 and moving two spaces to the right to position 3. Similarly, subtraction involves a leftward movement; subtracting the number two from the number one indicates the operation of starting at position 1 and moving two spaces to the left to position −1. Finally, we can think of multiplication as repeated movements along the number line, which are rightward, in the case of multiplication by a positive number and leftward, in the case of multiplication by a negative number. For example, 2×3 can be thought of as two successive rightward "jumps" of three units starting from the origin to reach the value 6. Similarly, 3×-1 can be thought of as three successive leftwards "jumps" from the origin, to give a value of −3.

The number line provides an important visual tool for understanding the relationship between integers and the arithmetic operations of addition, multiplication, and subtraction. However, it is also important because it allows us to define and understand more general definitions of numbers. We have defined the integers as evenly spaced points on the line, but is there a meaningful interpretation of the points which lie between the integers? One possibility is to interpret these points as *fractions* or, more formally, as *rational numbers*. Fractions can be thought of as points on the line, which can be expressed as the ratio of two integer numbers. For example, the point lying halfway between zero and one can be defined as 1/2, and the point lying halfway between zero and −1 can be defined as −1/2. We define the set of rational numbers as all numbers which can be written in the form a/b where a and b are integers with no common factors.[1] Alternatively, we can define rational numbers as those numbers which are solutions to equations of the form $bx-a=0$, where b and a are integers with no common factors. Note this definition is only meaningful when $b\neq0$. The set of rational numbers is written as $\mathbb{Q}$ and both the set of natural numbers and the set of integers are proper subsets of the set of rational numbers.

The rational numbers can be written in the form of fractions or as *decimal numbers*. Decimal numbers are written as a sequence of digits with a single separator referred to as the decimal point. For example, we can write $1/2=0.5$ or $1/4=0.25$. Not all decimal representations of rational numbers will have a finite number of digits or "decimal places." An obvious example here is the rational number $1/3$. If we divide one by three using the standard methods of division, then there will always be a remainder. We can write the

[1] An integer a is said to have *factors* c and d if $a=c\times d$, where c and d are both integers. The integers a and b are said to have a *common factor* c if they can be written in the form $a=c\times d$ and $b=c\times e$, where e is also an integer.

results of this calculation as $1/3 = 0.3333\ldots$, the ellipsis here indicates that this sequence will continue forever. An alternative notation for this is $0.\overline{3}$, which indicates a sequence of threes which continues infinitely. If a number is rational and has an infinite decimal representation, then it can be shown that the pattern of numbers in the expansion eventually repeats, for example, $7/11 = 0.636363\ldots = 0.\overline{63}$.

The number of possible rational numbers between any two points on the number line is infinite. For example, consider the two points, 0 and 1. Averaging these two values gives the rational number $1/2$ which lies halfway between these points. Now consider the interval defined by the numbers 0 and $1/2$, the rational number which lies halfway between these is $1/4$. We can now divide by two again to get the rational number which lies halfway between zero $1/4$ to get $1/8$, and there is no limit to the number of times we can do this. We can continue to define rational numbers using smaller and smaller intervals on the number line, and however small we make this interval, it can always be subdivided further by dividing it into two smaller subintervals.

Since all the rational numbers can be represented as points on the number line, and an infinite number of rational numbers lie between any two points on the line, it is tempting to think that any point on the line can be represented as a rational number. However, this is not true. To illustrate this, we will make use of a counter-example. Consider the equation $x = \sqrt{2}$. This states that x is equal to the square root of two. To find x, we look for a number which, when multiplied by itself, gives the integer value 2. However, it is not possible to find a rational number with this property. This can be demonstrated by the method of proof by contradiction. That is, we assume that the statement is true and then show that it implies a logical contradiction. If $x = \sqrt{2}$ is a rational number, then we should be able to find integers a and b (with no common factors) such that $(a/b)^2 = 2$. If this statement is true, then we have

$$\frac{a^2}{b^2} = 2 \Rightarrow a^2 = 2b^2. \tag{1.3}$$

It, therefore, follows that a is even. Let us write $a = 2k$ where k is an integer. From our definition of b, we have

$$4k^2 = 2b^2 \Rightarrow b^2 = 2k^2. \tag{1.4}$$

It, therefore, follows that b must also be even. The number 2 is, therefore, a common factor for both integers a and b, which contradicts the original assumption they have no common factors. Therefore, it is not possible to write $\sqrt{2}$ as a rational number. However, it is possible to write down an approximation to $\sqrt{2}$ using decimal notation as 1.41421..., but this decimal representation has an infinite number of terms and never settles down into a repeating pattern. Numbers without repeating patterns are referred to as *irrational numbers*.

Note that all numbers that can be written as a finite decimal expression are, by definition, rational. This should be immediately obvious. For example, suppose we have $x = 0.1234$, then we can equivalently write this as $x = 1{,}234 / 10{,}000 = 617 / 5000$. As we have already noted, however, not all numbers with infinite decimal expressions are irrational, for example, $1/6 = 0.16666\ldots$ where the number 6 repeats indefinitely. For a number to be irrational, it must have an infinite decimal expression that never repeats. Some of the most important numbers in mathematics fall into this category. Two examples are π, the ratio of the circumference of a circle to its diameter, and Euler's number *e*, which is defined as $\sum_{n=1}^{\infty} 1/n!$. These are both irrational numbers that have infinite, nonrepeating decimal representations. However, both can be represented by approximations. We have $\pi = 3.1416$ and $e = 2.7183$ to an accuracy of four decimal places.

We define the set of *real numbers* as the set of all numbers which can be written as infinite decimal expressions. This includes all the natural numbers, integers, rational numbers, and all those numbers that can be written as infinite, nonrepeating decimal expressions. The symbol for this set is $\mathbb{R}$ and, since all elements of this set can be thought of as points on the number line, it is usual to refer to this line as the *real line*. The set of real numbers is closed under addition, subtraction, and multiplication. That is, if a and b are real numbers, then $a+b$, $a-b$, and ab will also be real numbers. Moreover, it is closed under the operation of division, if we exclude the special case of division by zero. That is, if a and b are real numbers, then a/b is also real number except for the case $b = 0$.

The set of real numbers is the most general set we have defined so far because all previously defined number sets are subsets of this set. To clarify, we define a hierarchy of sets as shown in (1.5). This indicates that the set of natural numbers is a proper subset of the set of integers, which is a proper subset of the set of rational numbers, and which, in turn, is a proper subset of the set of real numbers. It follows that, if we can demonstrate a particular mathematical result is true for all real numbers, it will also be true when applied to natural numbers, integers, and rational numbers.

$$\mathbb{N} \subset \mathbb{Z} \subset \mathbb{Q} \subset \mathbb{R}. \tag{1.5}$$

REVIEW EXERCISES – SECTION 1.1

1. To which sets do the following numbers belong

(a) 0.25

(b) $2\sqrt{2}$

(c) –4

(d) 0.666….

(e) 5,489,127

2. Show that $\sqrt{36/25}$ is a rational number.

3. Show that $\sqrt{8}$ is irrational.

1.2 RULES OF ALGEBRA

> The rules of algebra provide a consistent method for the manipulation of symbols representing numbers. It is important to familiarize yourself with these rules because you will frequently need to use them.

Algebra is the mathematics of *symbols*. The use of symbols to replace numbers allows us to derive general rules which apply to all numbers within a given set. In this section, we apply the method of algebra to the four basic mathematical operations: addition, multiplication, subtraction, and division. Since the real numbers are the most general set of numbers we have defined so far, we will consider operations involving these numbers.

Commutative Property

The *property of commutativity* is concerned with the ordering of the variables in algebraic expressions. It states that, when performing addition or multiplication, the order of the variables is not important. Commutativity holds for the addition and multiplication of real numbers but not for subtraction and division. Let a and b be real numbers, and we can define the commutative properties as follows:

Commutative law of addition $\qquad a+b=b+a$

Commutative law of multiplication $\qquad ab=ba$

Note that the commutative property does *not* hold for either subtraction or division. This can easily be demonstrated using counterexamples.

Associative Property

The *property of associativity* concerns the grouping of operations. Parentheses are used to indicate the order of operations by grouping together those operations which are to be performed first. For addition and multiplication, the associativity property states that the order in which operations are carried out does not affect the result. We can show that the following rules apply for all real numbers a, b, and c:

Associative law of addition	$(a+b)+c=a+(b+c)$
Associative law of multiplication	$a(bc)=(ab)c$

Again, this property does not hold for subtraction and division.

Distributive Property

Distributivity is a property that applies when addition and multiplication form part of the same expression. It can be written as follows:

Distributive law of multiplication $\quad a(b+c)=ab+ac$

The distributive law states that, when evaluating a multiple of the sum of elements, we can either perform the summation first and then multiply by the common factor, or we multiply each of the elements by the common factor and then take the sum. Note that, unlike the commutative and associative laws, the distributive law does apply to the combination of multiplication and subtraction. In general, it is true that $a(b-c)=ab-ac$. It also applies to the combination of division with either addition or subtraction, i.e., $(b+c)/a=b/a+c/a$ and $(b-c)/a=b/a-c/a$, assuming that $a \neq 0$.

The properties of commutativity, associativity, and distributivity are fundamental to algebraic manipulation. If we carefully apply these rules, we can manipulate general expressions involving algebraic symbols to present them in more convenient forms. Although algebraic manipulation involves using only a few simple rules, it nevertheless requires practice to do this accurately and fluently.

Finally, we note that algebra also makes use of the existence of *additive and multiplicative identities* in the set of real numbers. The additive identity is the number 0, which has the property that $a+0=a$. Related to this idea, there exists an additive inverse $(-a)$ such that $a+(-a)=0$. The multiplicative identity is the number 1 which has the property that $a\times 1=a$. A related property is the existence of a multiplicative inverse $(1/a)$ such that $a\times(1/a)=1$. Note that the multiplicative inverse is only defined if $a \neq 0$.

Mathematical expressions can often involve multiple operations. For example, we might have an expression of the form $(a^2+b)c-d$. The value of this expression is sensitive to the order in which these operations are carried out. It is, therefore, important to establish rules for the precedence of different operations. The convention is to give priority to operations in parentheses, followed by exponents (or power) operations, followed by division and multiplication, and finally, addition and subtraction. In the United States, this is associated with the mnemonic *PEDMAS* or parentheses, exponents, division/multiplication, addition/subtraction. In the UK, the equivalent mnemonic is *BIDMAS* or brackets, indices, division/multiplication, and addition/subtraction. These mnemonics are not, however, completely unambiguous. The rules define "levels" for different operations, with parentheses being the top level, followed by exponents, then division/multiplication, and finally, addition/subtraction. However, when operations of the same level are written as part of the expression, then the application of alternative orderings may give different results. For example, $a-b-c$ involves two subtraction operations. The ordering does not tell us which of these we should perform first, and we have already seen that $(a-b)-c \neq a-(b-c)$. In cases like this, the convention is to work from the left to right of the expression so that $a-b-c$ is evaluated by first calculating b from a, and then subtracting c from the result. The use of parentheses, however, provides an unambiguous ordering for the operations and is recommended whenever the possibility of misinterpretation arises.

Some other notation conventions which are often assumed without being formally stated are:

1. The multiplication operator is often implicit. For example, $3 \times a$ is often written as $3a$.
2. Division is most often indicated by a horizontal line or slash rather than the division operator. That is, we write a/b rather than $a \div b$.

Table 1.2 gives a few illustrative examples of how the rules of algebra should be applied in practice.

TABLE 1.2 Evaluation order for algebraic expressions.

Example	Expression	Evaluation
1	$5a^4+2$	Raise a to the fourth power, multiply by 5, then add 2.
2	$(a+2)^3/b$	Add two to a, raise to the third power, and then divide by b.
3	$\sqrt{(2a+2)}$	Multiply a by 2, add 2, and then take the square root.

REVIEW EXERCISES – SECTION 1.2

1. Evaluate the following expressions.

(a) $4-(3-2)/3$

(b) $2(3-4)$

(c) $2/3-(3+1)4$

(d) $5\times 4-3/2$

(e) $6\div 3(1+2)$

2. Remove the parentheses from the following expressions, where a, b, and c are nonzero real numbers.

(a) $a(b+a)$

(b) $c-(a+b)$

(c) $c+(a-b)/c$

(d) $ca(a/c+b)-c$

(e) $a/b-(c+b)/c$

1.3 COMPLEX NUMBERS AND HYPERREAL NUMBERS

It is sometimes useful to extend the set of numbers we consider to include *complex numbers* (which include the square roots of negative numbers) and *hyperreal numbers* (which can be either smaller or larger than any of the set of real numbers.) In this section, we discuss both types of numbers and show how they are related to real numbers. However, we do not make immediate use of either set, so you can safely pass over this section and return to it later if you prefer.

Complex Numbers

The *algebraic real numbers* consist of the set of numbers that can be written as infinite decimal expansions and which are solutions to algebraic equations with integer coefficients. For example, the equation $x^2=2$ has solutions $x=\sqrt{2}$ and $-\sqrt{2}$ which are both algebraic real numbers. However, not all algebraic equations have real solutions. Consider, for example, the equation

$x^2 + 1 = 0$. We need to find a solution such that $x^2 = -1$, but since a squared real number is always positive, it follows that this equation has no real solution. This is unfortunate because equations like this occur naturally in all sorts of problems. To get around this problem, we define a new class of number known as *complex* or *imaginary* numbers. Let us define the symbol i to mean the square root of minus one, that is, $i = \sqrt{-1}$ and let a and b be real numbers. We can define the set of complex numbers as all numbers which can be written in the form $a + bi$.

A complex number of the form $x = a + bi$ consists of two parts, a real part a, and an imaginary or "complex" part $bi = b\sqrt{-1}$. The set of complex numbers is written $\mathbb{C}$ and consists of all numbers that can be written in this form. Note that the set of complex numbers includes the set of real numbers as a proper subset, $\mathbb{C} \supset \mathbb{R}$, because any real number can be written as a complex number with $b = 0$.

We can add, subtract, multiply, and divide complex numbers in the same way as we perform these operations for real numbers. To add two complex numbers together, we simply add the coefficients for the real parts and the complex parts, as shown in equation (1.6)

$$(a + bi) + (c + di) = (a + c) + (b + d)i. \tag{1.6}$$

EXAMPLE

Let $x = 1 + 2i$ and $y = 3 - 3i$, adding these numbers gives us $x + y = 4 - i$.

Subtraction of complex numbers operates by subtracting the corresponding coefficients, as shown in equation (1.7)

$$(a + bi) - (c + di) = (a - c) + (b - d)i. \tag{1.7}$$

EXAMPLE

Let $x = 4 - 2i$ and $y = -2 + 2i$, subtracting y from x gives $x - y = 6 - 4i$.

Multiplication of complex numbers is a little bit more complicated and requires the use of the distributive property of algebra. Using the property that $i^2 = -1$ gives us

$$\begin{aligned}(a+bi)(c+di) &= a(c+di)+bi(c+di)\\ &= ac+adi+bci+(bd)i^2\\ &= (ac-bd)+(ad+bc)i.\end{aligned} \tag{1.8}$$

EXAMPLE

Let $x=2+i$ and $y=1-i$, multiplying x by y gives $xy=3-i$.

Note that, if x and y are *complex conjugates*, that is, if $x=a+bi$ and $y=a-bi$, then their product is a real number. We can show this, in general, using the distributive property of multiplication since

$$\begin{aligned}(a+bi)(a-bi) &= a^2-(ab)i+(ab)i-b^2i^2\\ &= a^2+b^2.\end{aligned} \tag{1.9}$$

EXAMPLE

Let $x=3+2i$ and $y=3-2i$. The product of these two numbers is $xy=3^2+2^2=13$.

Finally, we can divide one complex number by another using the following procedure:

$$\frac{a+bi}{c+di}=\frac{(ac+bd)+(bc-ad)i}{c^2+d^2}. \tag{1.10}$$

The proof of this statement is left as Exercise 1.3.3 for the interested reader.

EXAMPLE

Let $x=1+2i$ and $y=2-2i$, using equation (1.10) we can show that $x/y=-1/4+(3/4)i$.

Earlier, we found that the real line provides a useful visual tool for understanding the nature of real numbers. In the case of complex numbers, a similar visualization is provided by thinking of them in terms of points in a two-dimensional plane. This is illustrated in Figure 1.3. The distance along the horizontal axis represents the real part of the complex number, and the distance along the vertical axis represents the imaginary or complex part.

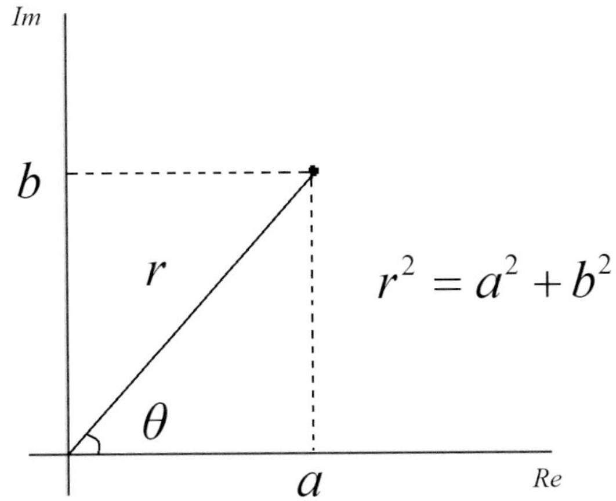

FIGURE 1.3 Diagrammatic representation of the complex numbers.

Points in a two-dimensional space can be represented in terms of their (x,y) coordinates, as shown in Figure 1.3, with the coefficient for the real part represented on the horizontal axis and that for the complex part represented on the vertical axis. This representation leads naturally to an alternative interpretation of complex numbers in terms of *polar coordinates.* Polar coordinates consist of the *magnitude*, i.e., the distance of the point from the origin and the *angle* of the point relative to the horizontal axis. These are represented by the symbols r and θ in Figure 1.3. The relationship between the two representations can be defined by the following pair of equations:

$$\begin{aligned} a &= r\cos\theta \\ b &= r\sin\theta\,. \end{aligned} \tag{1.11}$$

Polar coordinates prove useful when we use complex numbers to capture periodic motion, that is motion repeated in equal intervals of time. Let us consider how the location of a point in the (x,y) plane changes as the angle parameter changes while keeping the magnitude constant. The constant magnitude means that the length of the line from the origin to the point remains constant. Therefore, changing the angle over the range 0 to 2π has the effect of tracing out a circle in the plane, as shown in Figure 1.4. This means that complex numbers can be used to describe cyclical or periodic motion. In economic analysis, this proves useful when modeling phenomena such as business cycles.

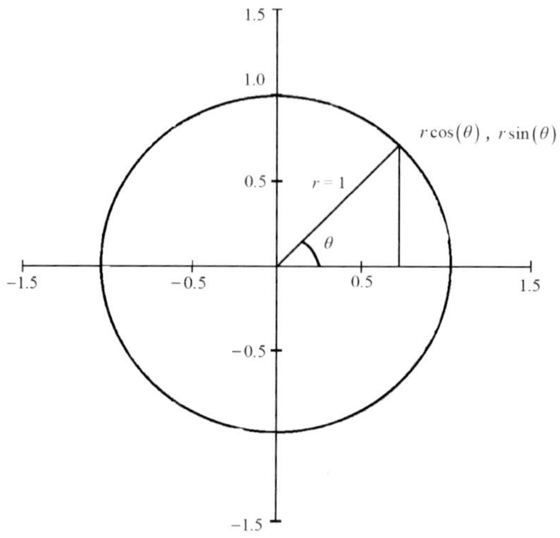

FIGURE 1.4 Effects of varying the θ parameter.

Hyperreal Numbers

Next, we turn to the set of *hyperreal numbers*. This extends the set of real numbers in two ways. First, to include extremely small numbers, or *infinitesimals*, and second, to include extremely large numbers, or *infinite* quantities. We introduce a discussion of these numbers here because we make use of them later in developing a treatment of calculus which is somewhat easier than the standard approach.

For many years, the use of infinitesimals in mathematics was regarded as lacking rigor. Many argued that they could not be defined clearly in the way that the real numbers are defined. In the 1960s, however, Abraham Robinson showed that infinitesimals and infinite numbers could be given rigorous mathematical definitions. This meant that the intuitive approach to the development of calculus used by Leibniz and Newton was retrospectively justified by modern mathematics. The number system that allows us to do this is referred to as the set of *hyperreal numbers*, and the approach to mathematical analysis which uses these numbers is referred to as *nonstandard analysis*. This distinguishes nonstandard analysis from *standard analysis*, which derives from the work of Weierstrass, which builds calculus using the method of limits.

There are three main principles of nonstandard analysis, which we set out below. Note that this is not meant to be a rigorous definition of the approach, but rather an intuitive introduction to the system that will allow us to make use of the concept of infinitesimals for the development of calculus in later chapters.

Principle 1: The Extension Principle

The set of real numbers is a proper subset of the set of hyperreal numbers. There exists at least one hyperreal number that is greater than zero but less than every positive real number. Formally, we state that there exists a nonzero number ε such that $-a < \varepsilon < a$, where a is any real number. ε (epsilon) is referred to as an *infinitesimal* number. The inverse of an infinitesimal number $H = 1/\varepsilon$ is an *infinite* number that is greater or less than any real number depending on whether ε is positive or negative. A hyperreal number is said to be *finite* if it lies between any two real numbers. Such numbers consist of a real number plus an infinitesimal. Finally, any function *f*, which is defined using real numbers, has a natural extension that can be applied to hyperreal numbers. For example, if $y = f(x) = 1 + 2x$ is a function that is defined for real numbers x and y, then there exists an extension of this function $f^*(x) = 1 + 2x$ that applies when x is a hyperreal number.

Principle 2: The Transfer Principle

Every real statement that holds for one or more real functions holds for the hyperreal natural extensions of these functions. This states that hyperreal numbers obey the same rules of arithmetic and algebra as the real numbers. For example, $a + b = b + a$ is true, whether a and b are real numbers or hyperreal numbers. Conventional functions and operations such as addition, subtraction, multiplication, etc., which are applicable to real numbers, can also be applied to hyperreal numbers. Note that the transfer principle, in conjunction with the extension principle, implies that there are many infinitesimal numbers. By the extension principle ε is infinitesimal, and, by the transfer principle, we can define the variable $\delta = 2\varepsilon$, which is also infinitesimal. Thus, since any multiple of an infinitesimal is itself infinitesimal, this implies that there exists an infinite number of infinitesimals.

Principle 3: The Standard Part Principle

Every finite hyperreal number is infinitely close to exactly one real number. This means that we can define any hyperreal number as the sum of a *standard* (or real) part and an infinitesimal. Thus, if a is a hyperreal number, then $a = \mathrm{st}(a) + \varepsilon$, where $\mathrm{st}(a)$ is real and ε is infinitesimal. Using these principles, we can write down the following rules for manipulating expressions that contain hyperreal numbers.

Rules for infinitesimal numbers

Let ε and δ be positive infinitesimal numbers, and let a be a nonzero real number.

1. Sum Rule	$\varepsilon + \delta$ is infinitesimal.
	$a + \varepsilon$ is finite but not infinitesimal.
2. Product Rule	$\varepsilon\delta$ and $a\varepsilon$ are both infinitesimal.
3. Quotient Rule	ε / a is infinitesimal.
4. Roots Rule	$\sqrt[n]{\varepsilon}$ is infinitesimal where $n \geq 1$.

In addition, we have the following rules for infinite numbers, where ε is a positive infinitesimal number and a is a nonzero real number.

Rules for infinite numbers

1. Reciprocals Rule	$H = a / \varepsilon$ is infinite.
2. Product Rule	aH is infinite.
3. Quotient Rule	ε / H is infinitesimal.
	H / ε is infinite.

One immediate implication of the reciprocals rule is that there is no unique number in the set of hyperreal numbers which we can refer to as "infinity." Instead, there are many infinite numbers depending on the values of a and ε, which we use to define H.

Note that these rules do not allow us to determine the nature of the product of an infinitesimal number and an infinite number which may be finite, infinitesimal, or infinite. Similarly, there are no definitive rules for the ratio of infinitesimal numbers, the ratio of infinite numbers, or the sum/difference of infinite numbers. Despite these limitations, however, the rules we have established will prove sufficient for us to derive all of the standard results of calculus using the method of nonstandard analysis.

REVIEW EXERCISES – SECTION 1.3

1. Show that the solutions of the equation $x^2 - 4 = 0$ are real, but those of $x^2 + 4 = 0$ are complex. Plot the curves $y = x^2 - 4$ and $y = x^2 + 4$ in the (x, y) plane and identify what makes them different.

2. Let ε be a nonzero infinitesimal number and a be a nonzero real number. For the following expressions, state the type of number

(a) $1/\varepsilon$

(b) $a+\varepsilon$

(c) $(a+1)\varepsilon$

(d) $1/(a+\varepsilon)$

(f) $(a+\varepsilon)/\varepsilon$

3. Let $x=a+bi$ and $y=c+di$, show that $\frac{x}{y}=\frac{(ac+bd)+(bc-ad)i}{c^2+d^2}$.

1.4 INTERVALS

An *interval* is a subset of one of the general sets of numbers we have defined. In this section, we introduce the idea of *open intervals* and *closed intervals* for real numbers. This will prepare the way for a discussion of functions in Chapter 2.

An *interval* defines a range of possible values that a number can take. We are often interested in intervals that define a range of possible values on the real line. For example, if a and b are real numbers, then the *open interval* (a,b) can be read as "the set of all real numbers which are greater than a but less than b." Open intervals are indicated by curved parentheses and do not include the end points. A *closed interval* is indicated using square parentheses, that is, $[a,b]$. This can be read as "the set of all real numbers greater than or equal to a but less than or equal to b." We can also define *semi-open intervals*, which mix these two definitions. For example, $[a, b)$ is the set of real numbers that is greater than or equal to a but less than b. All these intervals define subsets of the set of real numbers.

Intervals can be written in various ways. Table 1.3 shows some of these ways and defines some important cases. The precise definition of ranges becomes important in Chapter 2 when we consider the definition of functions. Ranges allow us to define the values of x over interval in which a function is valid, that is, its *domain*. For example, we may wish to restrict attention to numbers that lie in the range -1 to $+1$. We could indicate this by the open interval $(-1,1)$.

TABLE 1.3 Types of Interval and Methods for Defining Intervals

	Verbal definition	Set definition	Interval definition
Closed Interval	All real numbers which lie between the limits 1 and −1 including these values.	$S=\{x\in\mathbb{R}\mid -1\le x\le 1\}$	$[-1,1]$
Semi-Open Interval	The set of nonnegative real numbers.	$\mathbb{R}_{\ge}=\{x\in\mathbb{R}\mid x\ge 0\}$	$[0,\infty)$
Open Interval	The set of negative real numbers.	$\mathbb{R}^{-}=\{x\in\mathbb{R}\mid x<0\}$	$(-\infty,0)$

Intervals also provide a way of interpreting approximations of numbers that have an infinite decimal representation. We have already noted that some rational numbers have decimal representations which extend indefinitely. For example, we can state that 1/3 lies in the open interval $(0.332, 0.334)$. This is illustrated in Figure 1.5, which shows the location of 1/3 on the real line. The mid-point of this line, 0.333, is said to be accurate to three decimal places. Similarly, we have irrational numbers such as π which have infinite, nonrepeating decimal expansions. However, we can show that the value of π lies in the open interval $(3.1415, 3.1416)$, that is, we know that $3.1415<\pi<3.1416$.

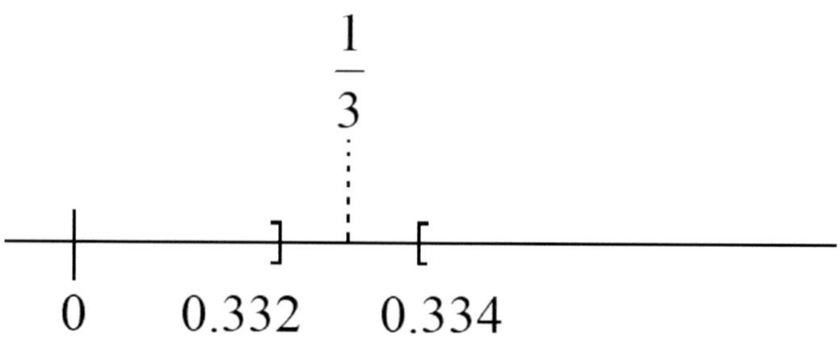

FIGURE 1.5 Representation of 1/3 on the real line.

In some circumstances, we may wish to use symbol ∞, which is used to denote infinity, as one of the limits of the interval. This symbol is used to indicate that the number can take on arbitrarily large positive or negative values. For example, if x lies in the interval $(-\infty,\infty)$, then this is simply equivalent to saying that it lies somewhere on the real line, that is, $x\in\mathbb{R}$. Alternatively, if x lies in the interval $[0,\infty)$ then this is equivalent to saying that x is a nonnegative real number. Note that we cannot have an interval that is closed at infinity (or minus infinity) because this symbol does not indicate a number in the conventional sense. Instead, it simply indicates that the variable concerned can be arbitrarily large in the case of ∞, or arbitrarily large and negative in the case of $-\infty$. Intervals like this, arise very frequently in the analysis of functions and have a particular notation for the relevant sets. For example, $\mathbb{R}_{>0}$ is used to indicate the set of real numbers greater than zero or $0 < x < \infty$. Similarly, $\mathbb{R}_{<0}$ is the set of real numbers less than zero, while $\mathbb{R}_{\geq 0}$ and $\mathbb{R}_{\leq 0}$ are the sets of real numbers greater than or equal to zero and less than or equal to zero, respectively.

REVIEW EXERCISES – SECTION 1.4

1. Use the following mathematical definitions to write a short English sentence that gives the range of numbers defined.

(a) $x \in \mathbb{R}_{\geq 0}$

(b) $x \in \mathbb{R}_{<0}$

(c) $x \in \mathbb{R}; \quad -1 < x < 1$

(d) $S = \{x \in \mathbb{Z}; -2 \leq x \leq 2\}$

(e) $x \in \mathbb{R}; \quad 0 < x < \infty$

2. From the following definitions in English, write down a mathematical definition of the relevant set.

(a) The set of positive real numbers greater than one.

(b) The set of positive real numbers less than or equal to one.

(c) The set of integers less than minus two.

(d) The set of real numbers less than ten but greater than, or equal to, one.

(e) The set of integers greater than, or equal to, zero.

1.5 EXPANDING AND FACTORIZING MATHEMATICAL EXPRESSIONS

> In this section, we discuss the use of parentheses in mathematical expressions to express them in convenient forms. The removal of parentheses from an expression is referred to as *expansion*, and that of introducing them, is referred to as *factorization*.

Parentheses are used to group together terms in mathematical expressions. By doing this, we can often express complicated expressions in relatively simple terms. However, it is often necessary to eliminate the parentheses for the purposes of evaluation and manipulation. This process is known as *expansion*. Expansion involves the application of the distributive properties of addition and subtraction. For example, suppose x is a real number, and we have an expression of the form

$$(2x+5)(x+4). \tag{1.12}$$

Using the distributive law of multiplication, we can write this expression as

$$\begin{aligned} &2x(x+4)+5(x+4)=2x^2+8x+5x+20 \\ &=2x^2+13x+20. \end{aligned} \tag{1.13}$$

Thus, the product of two linear expressions gives a *quadratic expression* in the variable x. A quadratic expression is any expression that can be written in the form ax^2+bx+c where a, b, and c are parameters. Expansion is a straightforward but occasionally tedious process. If the expression is a multiple of three linear expressions in x, then the outcome will be a *cubic expression*, that is, an expression of the form ax^3+bx^2+cx+d, where a, b, c, and d are parameters. For example, suppose we have

$$(2x-3)(x+1)^2. \tag{1.14}$$

Expanding $(x+1)^2$ gives an expression of the form x^2+2x+1. Therefore, (1.14) can be written as

$$\begin{aligned} &(2x-3)(x^2+2x+1) \\ &=2x(x^2+2x+1)-3(x^2+2x+1) \\ &=2x^3+4x^2+2x-3x^2-6x-3 \\ &=2x^3+x^2-4x-3. \end{aligned} \tag{1.15}$$

Providing that you are careful, expansion simply involves the application of the laws of algebra and, therefore, does not require any new or special mathematical techniques. It does, however, require care, attention to detail, not to mention practice.

Factorization of a mathematical expression is the reverse operation of expansion. It involves taking a polynomial expression and writing it as the product of lower-order expressions. For example, suppose we have an expression of the form $2x^2 - 2x - 12$. We have already seen that a quadratic expression can be the result of taking the product of two linear expressions. Therefore, it should be possible to reverse the process and write a quadratic expression as the product of two separate linear expressions. For our example, we have

$$2x^2 - 2x - 12 = (2x + 4)(x - 3) \tag{1.16}$$

which is easily confirmed by expanding the right-hand side to recover the original expression. In simple cases like this, we can often factorize quadratic expressions by inspection. In more general cases, however, this is not possible, and we need to find methods for factorizing quadratic and higher-order polynomial expressions.

Let us begin with the case of a quadratic expression, that is, a polynomial in which the highest power of x is its square. We can write a general quadratic polynomial as $ax^2 + bx + c$, where a, b, and c are real numbers. These are the *parameters* of the problem. Parameters are treated as fixed for any given expression but can be varied to create new expressions. By solving the problem in terms of general parameters, we, therefore, solve all possible problems which can be written in this form. Our objective is to find r_1 and r_2 such that

$$ax^2 + bx + c = (ax + r_1)(x + r_2). \tag{1.17}$$

This is an example of a *nonmonic quadratic expression* since we allow the coefficient on x^2 to take on the arbitrary value a. In practice, it is much easier to solve monic expressions in which $a = 1$. This does not involve any loss of generality since

$$(ax + r_1)(x + r_2) = a\left(x + \frac{r_1}{a}\right)(x + r_2). \tag{1.18}$$

Therefore, if we can solve for the factorization of the monic expression obtained by dividing the first factor by the parameter a, then we can easily work backward to factorize the original expression.

Consider the quadratic equation $x^2 + bx + c$. We wish to find r_1 and r_2 such that $x^2 + bx + c = (x + r_1)(x + r_2)$. Expanding the right-hand side of this expression yields $x^2 + bx + c = x^2 + (r_1 + r_2)x + r_1 r_2$. Hence, we look for solutions that satisfy the conditions $(r_1 + r_2) = b$ and $r_1 r_2 = c$. There are three possibilities here:

(1) r_1 and r_2 are distinct real numbers,

(2) there is a single real solution r such that $r_1 = r_2 = r$, and

(3) r_1 and r_2 are a pair of complex conjugate numbers.

EXAMPLE

Factorize the expression $4x^2 - 6x - 4$.

To find the factors of this expression, we first divide through by 4 to obtain the expression $x^2 - (3/2)x - 1$. Next, we look for values r_1 and r_2 such that $r_1 + r_2 = -3/2$ and $r_1 r_2 = -1$. In this case, we can see that the values $r_1 = -2$ and $r_2 = 1/2$ satisfy these conditions. Hence, we can write the factorization of the transformed expression as

$$x^2 - \frac{3}{2}x - 1 = (x-2)\left(x + \frac{1}{2}\right).$$

For the factorization of the original expression, we can multiply either of the two factors by 4. Thus

$$(4x-8)\left(x+\frac{1}{2}\right) \text{ and } (x-2)(4x+2)$$

are both acceptable factorizations of this expression. To check if you have performed factorization correctly, simply multiply the expressions and see if the operation recovers the original expression. We get alternative expressions here because factorization is simply a method of finding the *roots* of the quadratic expression. The roots are the values of x such that $4x^2 - 6x - 4 = 0$. For either of the alternative factorizations, this yields $x = 2$ and $x = -1/2$ as the solution.

Factorization by inspection is not always possible, so it is useful to develop a method to deal with more general cases. Let us return to the general quadratic expression $ax^2 + bx + c$. We have seen that we can factorize this by solving for the roots, i.e., the values of x such that the expression is equal to zero. Now, there is a standard solution for this problem, and we can show that the values of x, which are consistent with $ax^2 + bx + c = 0$ are given by

$$x_{1,2} = \frac{-b \pm \sqrt{b^2 - 4ac}}{2a}. \qquad (1.19)$$

The derivation of this result is relatively straightforward but somewhat lengthy; therefore, it is simply stated here without proof. Note, however, that it immediately establishes the conditions in which different solutions are obtained. If $b^2 - 4ac > 0$, then we have real distinct roots, if $b^2 = 4ac$ then we have a single real root and if $b^2 - 4ac < 0$, then we have complex conjugate roots. The condition given in equation (1.19) gives us a general method for finding the roots of any quadratic expression and, hence, factorizing it.

EXAMPLE

Consider the expression $2x^2 - 7x + 3$. This has roots

$$x_{1,2} = \frac{7 \pm \sqrt{49 - 24}}{4}$$

$$x_1 = 3 \text{ and } x_2 = \frac{1}{2}.$$

We can therefore write the factorization of the expression as

$$\begin{aligned} &2(x-3)\left(x - \frac{1}{2}\right) \\ &= (2x-6)\left(x - \frac{1}{2}\right) \text{ or } (x-3)(2x-1). \end{aligned}$$

Again, you can easily check that these are both acceptable factorizations by expanding them to recover the original expression.

As the order of the expression (the highest power of x) increases, the number of roots increases, and it becomes harder to solve for these roots using the methods we have described for quadratics. Therefore, for factorization of higher-order polynomial expressions, we often need to use numerical methods to solve for the roots of an expression in order to factorize it.

A useful trick is that if we can find one root of the expression by inspection, then we can reduce the problem to one of lower order. For example, a cubic expression will have three roots. If we can find one of these immediately, then we can turn the problem into the simpler one of finding the roots of a quadratic expression. The following example illustrates this process.

EXAMPLE

Suppose we wish to factorize the cubic polynomial expression $4x^3 - 7x + 3$. By inspection, we note that $x = 1$ is a root since the value of the expression when $x = 1$ is zero. Hence, we can extract this factor from the expression and write it as

$$(x-1)\{ax^2 + bx + c\}.$$

We can determine the parameters of the quadratic expression in curly parentheses by expanding and equating coefficients. We have

$$(x-1)\{ax^2 + bx + c\} = ax^3 + (b-a)x^2 + (c-b)x - c.$$

Equating coefficients gives us $a = 4, b - a = 0$, and $c = -3$. Therefore, we need to factorize the quadratic expression $4x^2 + 4x - 3$ in order to find the two remaining roots. We have

$$x_{1,2} = \frac{-4 \pm \sqrt{16+48}}{8} = \frac{-4 \pm 8}{8}$$

$$x_1 = -\frac{3}{2} \text{ and } x_2 = \frac{1}{2}.$$

We have therefore solved for all three roots of the expression, and we, therefore, write it in the form

$$4\left(x + \frac{3}{2}\right)\left(x - \frac{1}{2}\right)(x-1).$$

REVIEW EXERCISES – SECTION 1.5

1. Expand the following expressions.

(a) $(x+1)(x+2)$

(b) $(2x+1)(x+3)$

(c) $(x+1)(x-1)$

(d) $(x+3)^2$

(e) $x+x(x-1)$

2. Factorize the following expressions.

(a) x^2+2x+1

(b) $9x^2+12x+4$

(c) $x^2+x+1/4$

(d) $2x^2+12x+18$

1.6 A NUMERICAL METHOD FOR FINDING ROOTS

In Section 1.5, we discussed methods for finding the roots of polynomial equations. It is often possible to find solutions for simple lower-order polynomial equations by inspection or by using the standard formula for quadratic equations. As the order of the polynomial increases, it becomes increasingly difficult to solve for its roots by these methods. However, numerical methods allow us to solve for the roots of polynomials in these more general cases. In this section, we illustrate the use of the *bracketing algorithm*, to calculate the roots of a cubic polynomial. This method can easily be applied to more general cases.

Consider the equation $x^3-2x=0$. Because this is a cubic equation, there will be up to three values of x that satisfy this relationship. An obvious solution is $x=0$, but how do we go about finding the other two? Well, we know that if $x=1$ then $x^3-2x=-1$, and we know that if $x=2$, then $x^3-2x=4$. Here, the value of the expression changes sign between these values. Assuming that the value of the expression changes continuously between these points, which is the case, it follows that it must be equal to zero at some intermediate point. We know, therefore, that there is a solution somewhere in the interval $(1,2)$. We can narrow this interval by examination of the intermediate point $x=1/2$. If the expression is negative at this point, then we can set this as the lower limit, and if it positive, then we can set this as the upper limit. In this case, $x=1.5$ gives $x^3-2x=0.375$. Hence, we know that the root lies somewhere in the interval $(1, 1.5)$ as illustrated in Figure 1.6.

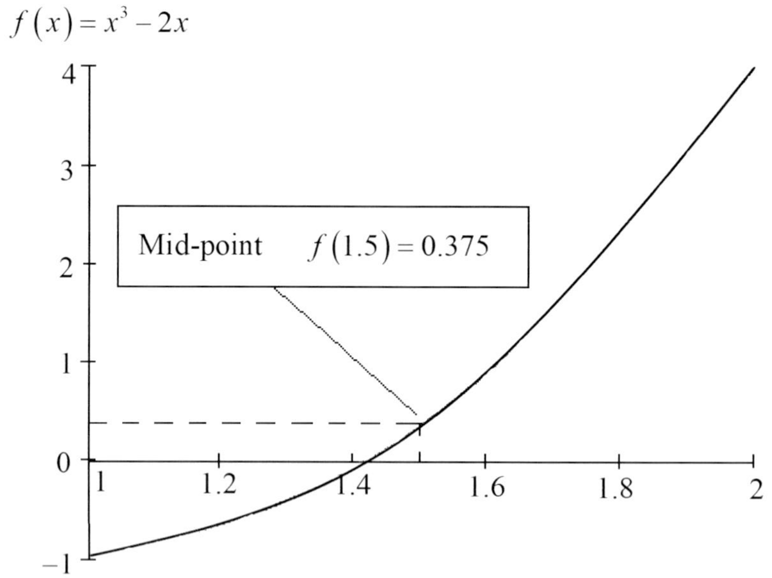

FIGURE 1.6 Interval estimate for root.

Having narrowed the interval once, we can repeat the procedure again with $x = 1.5$ as the new upper limit. In fact, we can continue to repeat this process until the lower and upper limits are sufficiently close to each other to judge that the solution has converged. This method is known as the *bracketing method* for finding the roots of equations, and it provides a robust algorithm for finding the roots of a polynomial equation, providing we can find an interval in which the expression changes sign and that it varies continuously along that interval.

Figure 1.7 gives Python code that implements the bracketing method for the equation $x^3 - 2x = 0$, starting with the interval $[1,2]$. When the tolerance level is set at 10^{-8}, that is, we require an answer which is accurate to seven decimal places, then we find a solution $x = 1.414213$. This gives us one of the nonzero roots of our equation. To find the other, we set the initial interval at $[-2,-1]$, then we can show that there is another solution at $x = -1.414213$. Finally, if we set the initial interval to $[-1,1]$, then we confirm numerically that there is a third solution at $x = 0$.

```
import sys

def f(x):
  f = x**3-4.72*x**2-3*x+14.16
  return f

xlower = 2
xupper = 5
tol = 10**-8

counter = 0

while abs(f(xupper))>tol:
 counter = counter+1

 if f(xlower)*f(xupper)>0:
  print("Function has same sign at lower and upper limits")
  print("Value at lower limit ",f(xlower))
  print("Value at upper limit ",f(xupper))
  sys.exit()

 if f(xlower)==0:
  print("There is a root at x = ",xlower)
  sys.exit()

 if f(xupper)==0:
  print("There is a root at x = ",xupper)
  sys.exit()

 xmid =(xlower+xupper)/2

 if f(xlower)<0 and f(xupper)>0:
   if f(xmid)<0:
     xlower=xmid
   if f(xmid)>=0:
     xupper = xmid

 if f(xlower)>0 and f(xupper)<0:
   if f(xmid)<=0:
     xupper = xmid
   if f(xmid)>0:
     xlower = xmid

 if counter>100:
    print("Convergence not achieved in 100 iterations")
    print("Root not found")
    sys.exit()

if counter<100:
  print ("There is a root at x = ",xupper)
```

FIGURE 1.7 Python algorithm for the bracketing method.

REVIEW EXERCISES SECTION 1.6

1. Modify the code in Figure 1.7 to solve for the root of the equation $x^3 - 3x = 0$, which lies in the interval $[-4, -1]$.
2. Modify the code in Figure 1.7 to solve for the root of the equation $x^3 - 2x^2 - 2x = 0$, which lies in the interval $[-1, -0.5]$.

CHAPTER 2

LINES, CURVES, FUNCTIONS, AND EQUATIONS

Functions take the elements of one set as an input and assign to them the elements of another set as the output. Relationships of this kind occur frequently in economic analysis. For example, the demand function defines the quantity of a good purchased as a function of its price. In this chapter, we develop the theory of functions and consider a variety of mathematical forms that are useful for economics and business analysis.

2.1 THE CARTESIAN PLANE

Mathematics is always easier if we can visualize the processes being described. In the case of functions, we can often present relationships in terms of lines or curves in two-dimensional space. Therefore, we begin our treatment of functions by developing the basic tools needed to present simple functions as graphs.

Diagrammatic representation is an important part of understanding functions. This is particularly useful for us because so much of economic theory is expressed in terms of geometric objects. An obvious example here remains the demand and supply curves taught to all students of economics at the beginning of their studies. The geometric ideas covered in this chapter are concerned with the two-dimensional surface known as the *Cartesian plane*. However, most of the ideas we cover will generalize easily to higher-order dimensions.

Imagine a flat sheet of graph paper that extends infinitely in all directions. You now have a good idea of what is meant by the Cartesian plane. A point in this plane is a location defined by two *coordinates*. These are distances from an arbitrary point known as the *origin*. Passing through the origin are a vertical line and a horizontal line which, by convention, are labeled the y-axis and the x-axis, respectively. The location of any point in the plane is defined by measurements along these axes and is referred to as the (x,y) coordinates of the point. This is illustrated in Figure 2.1.

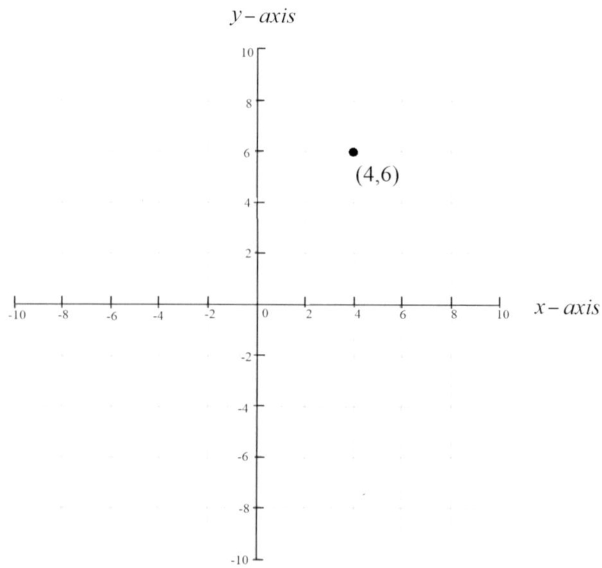

FIGURE 2.1 The Cartesian plane.

The study of geometry using the Cartesian plane is known as *Cartesian geometry*, or alternatively, as *coordinate* or *analytic geometry*. Three important objects in this type of geometry are the *point*, the *line*, and the *curve*. A point is defined by an *ordered pair* of coordinates (x,y) that measure the distance along the x and y axes, respectively. For example, the point $(4,6)$ is marked on the diagram. A *line* is defined as the set of points linking two points (x_1,y_1) and (x_2,y_2) where the gradient $b=(y_2-y_1)/(x_2-x_1)$ is constant. The equation of such a line is given by $y=a+bx$, where $a=y_1-bx_1$. A *curve* is a generalization of the line in that it also consists of a set of points linking two points, but, in this case, the slope is not necessarily constant. We can think of lines and curves as the paths followed by an object traveling from one point to another in the plane. The line provides the shortest such path, while a curve is a more general definition that allows for alternative paths.

A *Cartesian equation* is an equation that defines a curve in the Cartesian plane. Such equations are normally defined in terms of *parameters*, which are fixed for a particular curve but can be varied to create alternative curves with the same shape. For example, we have already defined the equation of a straight line as $y = a + bx$, where the parameters are the *intercept* (a) and the *slope* (b). By changing the values of these parameters, we can trace an infinite number of curves with the basic characteristics of a straight line. However, we are not restricted to linear relationships. Different equations can be used to trace out different kinds of curves in the plane. Some examples of equations that define curves in Cartesian space are given in Table 2.1, where a and b are parameters. The type of curves which are generated by these equations are illustrated in Figure 2.2.

TABLE 2.1 Cartesian equations for various curves.

Straight line	$y = a + bx$
Circle	$x^2 + y^2 = a^2$
Ellipse	$\frac{x^2}{a^2} + \frac{y^2}{b^2} = 1$
Hyperbola	$\frac{x^2}{a^2} - \frac{y^2}{b^2} = 1$
Parabola	$y^2 = 4ax$

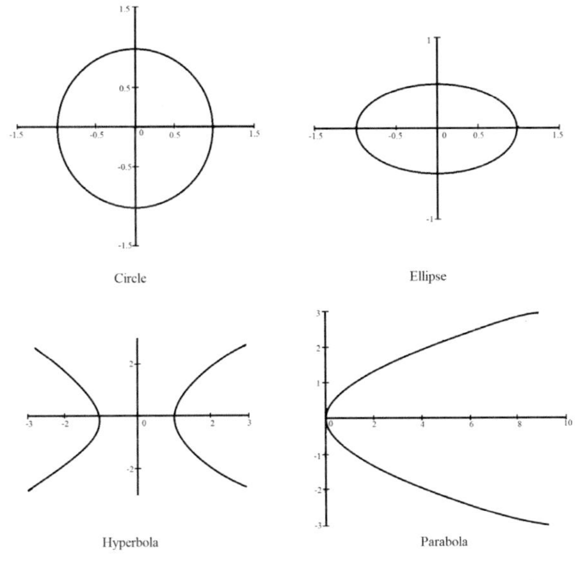

FIGURE 2.2 Cartesian equations in the Cartesian plane.

It is sometimes useful to define curves in *parametric form* rather than as an equation linking the x and y coordinates. When using parametric form, we write the x and y coordinates in terms of a third variable. For example, the parabola $y^2 = 4ax$ can be written in parametric form using a third variable t. This takes the form as $(x,y)=(at^2, 2at)$, where t can be any real number, and a is a parameter. Parametric form is often useful when describing motion through time. Hence, the symbol often used for the third variable is the letter t. The parametric forms of the circle, ellipse, and hyperbola are, respectively $(a\cos t, a\sin t)$, $(a\cos t, b\sin t)$, and $(a\sec t, b\tan t)$, where a and b are parameters, and the sine (sin), cosine (cos), secant (sec), and tangent (tan) functions are defined in Section 2.7.

EXAMPLE

If we vary the parameter of the parabola equation, then the result is a curve with the same general shape as the original curve but is displaced from it in some direction. Figure 2.3 compares parabola equations with parameters 1 and 2, respectively. As the parameter a increases, the curve retains the original shape but is more widely spread around the x-axis than the original curve.

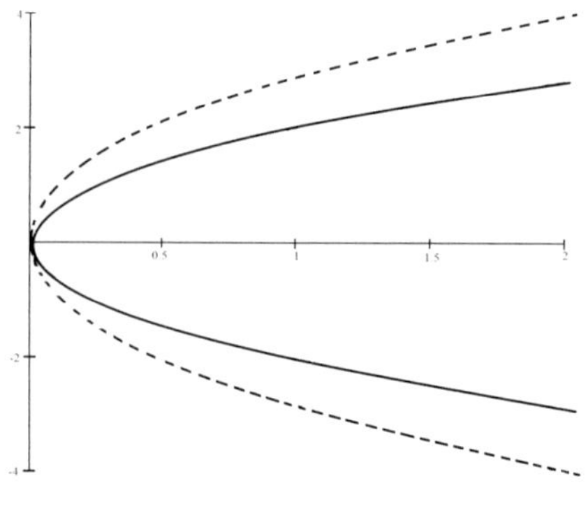

FIGURE 2.3 Parabolas with parameters 1 (solid line) and 2 (broken line).

REVIEW EXERCISES – SECTION 2.1

1. Calculate the equations of the straight lines passing through the following pairs of coordinates.

   (a) $(1,-1)$ and $(3,-5)$

   (b) $(1,7)$ and $(2,11)$

   (c) $(1,2)$ and $(4,11)$

2. Given the following equations for straight lines, calculate the values of x, which give $y = 0$.

   (a) $y = 4 + 3x$

   (b) $y = -1 + 2x$

   (c) $y = -3 + \frac{1}{2}x$

3. Variables x and y both vary with time according to the formulas $x = 4t - 1$ and $y = 3t$. This defines a curve in the Cartesian plane. Find the alternative representation in the form of an equation which links y and x.

2.2 FUNCTIONS

> A function is a rule which takes an element from one set and maps it to the elements of another set. Although functions are often written in the form of equations, they are not the same thing.

A *function* is a rule which associates objects in one set with objects in another set. For example, a function could be a rule which takes one number (the *argument* or *input*) and uses it to assign another number (the *output* or *result*). Equations are often used to define the rule, but simply writing down an equation relating two variables is not sufficient to define a function. To fully define a function, we must also specify the sets of numbers which are valid inputs and outputs of the relationship we define. A simple example is an equation of the form $y = x + 2$. For this to be a function, we must also specify the set of numbers from which x is drawn and the set of numbers that comprises the possible outcomes, y. These are referred to as the *domain* and the *codomain* of the function. For example, we can define a function using the relationships shown in (2.1):

$$\begin{aligned} &y = f(x) = x + 2 \\ &f : \mathbb{Z} \rightarrow \mathbb{Z} \end{aligned} \quad . \tag{2.1}$$

The first part of the definition consists of the equation $y = x + 2$. This defines the rule which takes x, the *argument* of the function, and maps it to y, the *output*. The second part of this function defines the domain and the codomain. In this example, we say that the function f "maps" the set of integers to the set of integers. The notation $f : \mathbb{Z} \rightarrow \mathbb{Z}$ can be read as "f maps the set of integers to itself." Note that the same equation could be used to map the set of real numbers to itself, that is, $f : \mathbb{R} \rightarrow \mathbb{R}$, but this would be a different function.

A function is defined as a *mapping* of elements in the domain to elements in the codomain. This is illustrated in Figure 2.4 which shows that every element in the set to the left is associated with an element in the set to the right. The set of values taken by the function is referred to as the *image* or the *range* of the function. The codomain and the range of the function may be different because there is no requirement that every element in the codomain have a matching item in the domain. This is illustrated in the diagram where there is no element in the domain associated with the element y_4 in the codomain. However, the range is always a subset of the codomain.

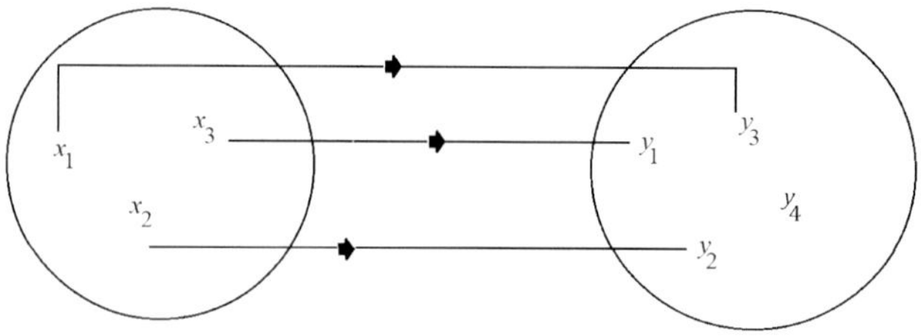

FIGURE 2.4 Functions and mapping.

Consider a curve in Cartesian space defined by an equation of the form $y = f(x)$. This may or may not be consistent with a functional relationship, depending on the definition of the domain. A condition for f to be a function is that each value of x in the domain must be associated with a unique value of y. This means that not all the curves discussed in the previous section can be thought of as functions unless we restrict the domain in some way. For a straight-line equation, this is not a problem because, for every value of x on the real line, there is a unique corresponding value of y. However, if we take the case of the parabola defined by $y^2 = 4ax$, and we define the domain as the set of nonnegative real numbers, then this does not define a function because, apart from $x = 0$, every value of x is associated with two different values of y. For example, if $x = 1$ then the curve is consistent with both $y = 2\sqrt{a}$ and $y = -2\sqrt{a}$.

In economic or business examples, we are most often concerned with functions where the domain and the codomain are subsets of the set of real numbers. It is often necessary to restrict the domain to values of x which are economically meaningful. For example, when considering demand and supply relationships, negative price and quantity values cannot occur. Suppose we are interested in the properties of a linear demand curve of the form $p(q) = a - bq$. In purely mathematical terms, it is possible to define the domain as the set of real numbers, but this allows for inputs and outputs which do not make economic sense. We can avoid this by defining the domain as the closed interval $[0, a/b]$ which gives the range as $[0, a]$. By defining the domain in this way, we ensure that the function does not imply negative price or output levels.

Linear functions are of particular interest to us because they often provide the simplest form in which we can approximate economic relationships. Before going on to consider more complex relationships, we will spend some time looking at the properties of linear functions. First, we note that the set of real numbers is a *closed set* under the operations of addition and multiplication. That is, the addition or multiplication of two real numbers always produces a real number as the output. Therefore, a linear function with a real input and real parameters will always produce a real output. This property is one of the reasons why linear relationships are particularly easy to work with.

Let us consider the properties of the linear function defined by

$$y = f(x) = a + bx \tag{2.2}$$

where both the domain and the codomain consist of the set of real numbers. The parameters in equation (2.2) are the *intercept* (a) and the *gradient* or *slope* (b). The intercept is the value of y when $x = 0$, and the slope is the change in y with respect to x, which is constant for a linear function. We can calculate the slope of a linear function by dividing the change in y by the change in x over any interval. That is, if we take any two points on the function (x_1, y_1) and (x_2, y_2), and calculate $\Delta y = y_2 - y_1$ and $\Delta x = x_2 - x_1$, then the gradient $\Delta y / \Delta x$ is the same, regardless of the choice of x_1 and x_2. The Δ (delta) notation is frequently used in mathematics to denote a discrete change in a quantity. That is, the change between two different points on a curve defined by an equation.

A quantity that is related to the gradient is the function's *elasticity*, the response of y to changes in x. This is defined as the proportional, or percentage, response of the y variable to a given proportional change in the x variable. Thus, in general, we can define the elasticity as

$$\eta = \frac{(\Delta y / y)}{(\Delta x / x)} = \frac{\Delta y}{\Delta x}\frac{x}{y}. \tag{2.3}$$

An important case is the *price elasticity of demand*. This measures the proportional change in quantity demanded resulting from a given proportional change in the price. In the case of a linear demand curve, the price elasticity of demand will be different at different points on the curve. Although $\Delta y / \Delta x$

is constant on a linear demand curve, the ratio x/y varies along the curve. Since, in most cases, we expect demand to respond negatively to price, there is a long-standing convention that this quantity is multiplied by minus one so that the price elasticity is expressed as a positive quantity. That is, we can define the price elasticity of demand as

$$\eta_P = -\frac{\Delta q}{\Delta p}.\frac{p}{q} \tag{2.4}$$

where p and q are price and quantity demanded.

EXAMPLE

Consider the linear demand curve $p(q) = a - bq$. The domain for this function can be defined as the closed interval $[0, a/b]$ since negative quantities are not possible and $q > a/b$ implies a negative price. The price elasticity of demand is defined as $\eta = -(\Delta q / \Delta p)(p/q)$ and, since $\Delta q / \Delta p = -1/b$, it follows that this depends on the ratio p/q. Substituting for $\Delta q / \Delta p$ gives us $\eta = (1/b)(p/q)$. This means that the elasticity can take on any value between 0 (when $p = 0$) and ∞ (when $q = 0$). This is illustrated in Figure 2.5.

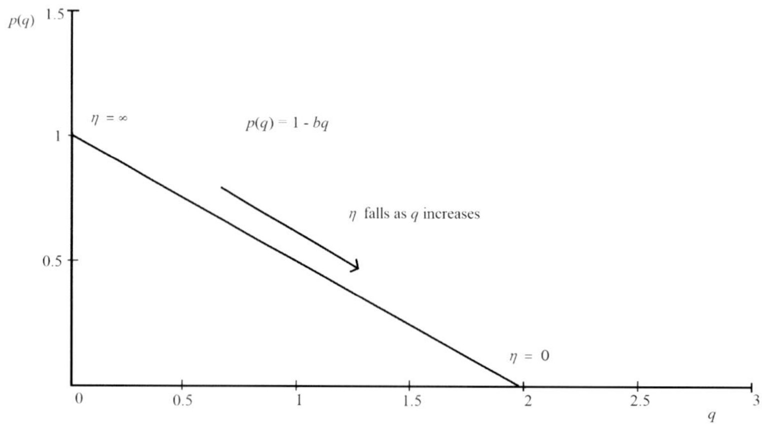

FIGURE 2.5 Properties of a linear demand curve.

Now, let us return to the general case and plot a linear function in the Cartesian plane. This gives us the kind of relationship shown in Figure 2.6, where, in this case, the intercept is equal to one, and the gradient is equal to 0.5.

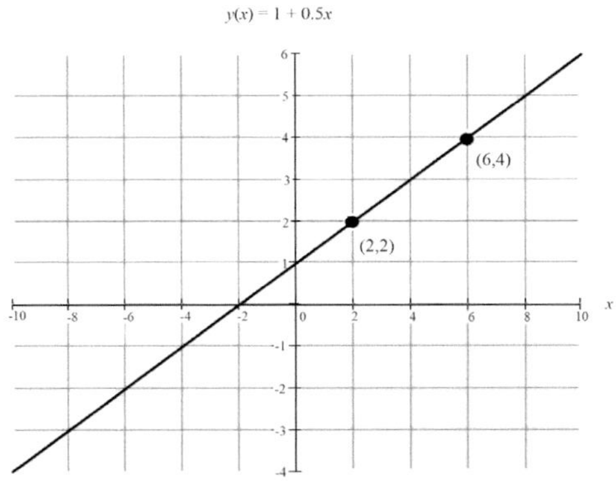

FIGURE 2.6 A linear function in the Cartesian plane.

Linear functions are particularly easy to draw because we can choose any two points on the curve and simply extend the straight line between them indefinitely. In Figure 2.6, we choose two points on the function $(2,2)$ and $(6,4)$, which then allows us to draw the complete function by simply extending the straight line between these points indefinitely to both the left and the right. The gradient is calculated as the change in y divided by the change in x

$$b = \frac{4-2}{6-2} = \frac{2}{4} = 0.5. \tag{2.5}$$

To calculate the intercept, we take either of the two points and calculate the value of a which is consistent with the slope we have already computed. For example, we know that a must be consistent with $(x,y)=(2,2)$ and therefore $2 = a + 0.5 \times 2$ which gives $a = 1$. This approach generalizes to the case where we are given any pair of points in the (x,y) plane. For any two points (x_1, y_1) and (x_2, y_2) the gradient and intercept are given by the formulas shown in equation (2.6).

$$b = \frac{y_2 - y_1}{x_2 - x_1} \qquad a = \frac{x_2 y_1 - x_1 y_2}{x_2 - x_1}. \tag{2.6}$$

The linear function is a *one-to-one* function. What this means is that every value of y in the range of the function is associated with a single value of x in the domain. Not all functions have this property. For example, consider the quadratic function $f(x) = x^2$ where the domain is the set of real numbers. For this function, we have $f(2) = f(-2) = 4$, and therefore, the quadratic function is not one-to-one. A sufficient condition for a function to be one-to-one is that it is *monotonic*. A monotonic function is a function whose slope never changes sign.

If a function is one-to-one and its range is equal to its codomain, then for every input value x, there is a unique output value, y, and vice-versa. This means that the function has an associated *inverse function* $x = g(y)$. For every possible value of y, we can find a unique value of x which is consistent with the original function. The inverse function is often written using the notation $x = f^{-1}(y)$.

EXAMPLE

Consider the linear function $y = f(x) = 1 + 2x$ where the input is a real number. This is one-to-one because each value of y is associated with a unique value of x. Moreover, we can find a value of x which will generate any real value y. It follows that this function has an inverse function which is given by the equation $x = -1/2 + y/2$, where both the original and inverse functions have domain and codomain equal to the set of real numbers.

The existence of an inverse function will depend on the definition of the domain. In some circumstances, we can ensure that a function has an inverse by restricting the domain appropriately. For example, consider the function $f(x) = x^2$ where x is the set of real numbers. This function is not one-to-one because $f(a) = f(-a)$ when $a \neq 0$ and, therefore, this function does not have an inverse function. However, if we define the domain as the set of nonnegative real numbers, then the resulting function becomes one-to-one, and an inverse function does exist. Therefore, for $y = f(x) = x^2$ where $f : \mathbb{R}_{\geq 0} \to \mathbb{R}_{\geq 0}$, the inverse function takes the form, $x = g(y) = \sqrt{y}$ where $g : \mathbb{R}_{\geq 0} \to \mathbb{R}_{\geq 0}$. This illustrates the importance of specifying the domain as well as the equation when defining a function.

REVIEW EXERCISES – SECTION 2.2

1. For each of the following functions, determine the maximum domain (i.e., the maximum interval on the real line for which y is defined) and the corresponding range for the function.

(a) $y = \sqrt[3]{x}$

(b) $y = 1/x^2$

(c) $y = |x|$

(d) $y = -3x^2$

2. Identify any real numbers x such that the following expressions are not defined.

(a) $1/(x-1)$

(b) $x/(2+x)$

(c) $3-1/(x^2-4)$

(d) $1/(x^3-8)$

3. Determine which of the following expressions defines a function if the domain is the set of real numbers.

(a) $|y| = x$

(b) $y = |x|$

(c) $y^2 - 2x = 0$

(d) $3y + 2x = 1$

2.3 LIMITS

Limits are an important mathematical tool in the development of calculus. A limit determines how the output value of the function behaves as the input approaches some particular value. Limits are often a stumbling block for nonmathematicians because they appear highly technical. However, the underlying ideas are often very simple and can be easily understood using a graphical approach.

A *limit* is the value toward which a function tends as the value of x gets close to, but not equal to, a particular value. We write limits using the notation $\lim_{x \to a} f(x)$. This can be interpreted as the value toward which $f(x)$ tends as x gets close to a. For simple functions, limits are often obvious. For example, suppose we have $f(x) = 2x$ where x is a real number. The limiting value of the function as x tends to the value 1 is simply equal to the value of the function at that point, that is $\lim_{x \to 1} 2x = 2$.

Limits become more interesting and harder to deal with when the function is more complicated. Consider the equation $y = f(x) = 1/x$. In this case, $1/x$ is not defined for $x = 0$ and, although it is defined for all other real numbers x, it behaves oddly for values of x close to zero. If x is positive but close to zero, then $f(x)$ is both large and positive. However, if x is negative and close to zero, then $f(x)$ is large and negative. This means that the function exhibits a *discontinuity* at this point, as shown in Figure 2.7. For this equation to be interpreted as a function, it is necessary to exclude zero from the domain.

In cases where the function exhibits a discontinuity, we need to make a distinction between *left limits*, or *limits from below*, and *right limits*, or *limits from above*. The left limit is the limit of $f(x)$ as x approaches some value a for values of $x < a$, while the right limit is the limit of $f(x)$ as x approaches a for values of $x > a$. We can write these are $\lim_{x\to a^-} f(x)$ and $\lim_{x\to a^+} f(x)$, respectively. Let us consider the example of $f(x) = 1/x$. For positive values of x, the value of $f(x)$ becomes very large as x gets close to zero. In terms of limits, the right limit of $f(x)$ is infinity. We can write this as $\lim_{x\to 0^+} (1/x) = \infty$. Similarly, for negative values of x, the value of $f(x)$ becomes large but negative as x gets close to zero. Alternatively, the left limit of $f(x)$ is equal to minus infinity, which can be written as $\lim_{x\to 0^-} (1/x) = -\infty$.

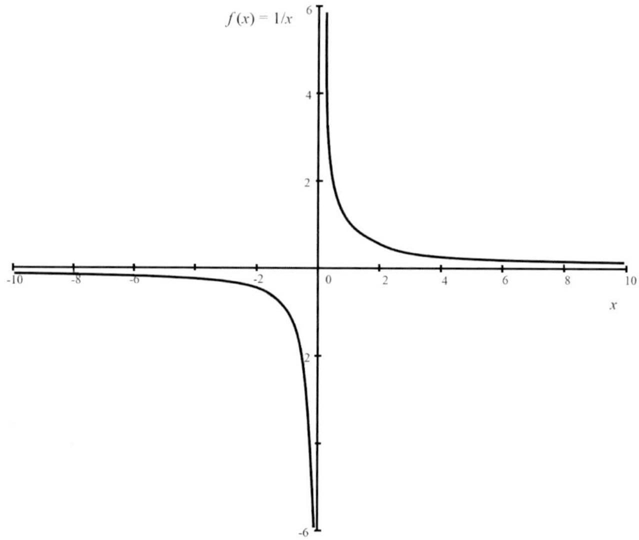

FIGURE 2.7 Plot of function $y = f(x) = 1/x$; $f:\mathbb{R}^* \to \mathbb{R}^*$.

Note that, by stating that a limit is equal to infinity, we do not mean that infinity can be treated as a number in the conventional sense. Rather, the value of the function becomes arbitrarily large as the value of x approaches its limiting value. Formally, we say that

$$\lim_{x \to a} f(x) = \infty \text{ if for any value of } L > 0 \text{ there exists a}$$
$$\text{a } \delta > 0 \text{ such that } |x - a| < \delta \Rightarrow f(x) > L.$$

In our example, the right limit of $1/x$ is infinite because we can make $1/x$ as large as we wish by choosing a value of x which is positive and sufficiently close to zero. Similarly, the left limit is equal to minus infinity because we can make $1/x$ as large and negative as we wish by choosing a value of x which is negative and sufficiently close to zero.

The example given here is for the important special case when the limit is infinite. The definition of limits is more general. We can state the general definition of a limit as follows

$$\lim_{x \to a} f(x) = L \text{ if for any number } \varepsilon > 0 \text{ there exists a number } \delta > 0$$
$$\text{such that } |x - a| < \delta \text{ ensures that } |f(x) - L| < \varepsilon$$

It follows that, by setting x close to some value a, we can ensure that $f(x)$ is as close as we wish to its limiting value L. Note that, for a limit to exist according to this definition, the left and right limits must be equal. That is, we must have $\lim_{x \to a^-} f(x) = \lim_{x \to a^+} f(x) = L$. If the left and right limits of the function are equal, then we say that the function is *continuous* at this point. If this is not true, then we say that the function is *discontinuous* at this point. A *continuous function* is defined as a function for which $\lim_{x \to a} f(x) = f(a)$ for all values of a in the domain. The function $f(x) = 1/x$ is said to have a *discontinuity* at $x = 0$.

Note that we can modify the definition of limits to include the case when x tends to infinity. In this case, we have

$$\lim_{x \to \infty} f(x) = L \text{ if for any number } \varepsilon > 0 \text{ there exists a number } a$$
$$\text{such that } x > a \text{ ensures that } |f(x) - L| < \varepsilon$$

This definition becomes important when wish to define the *asymptotes* of a curve or function. The asymptotes of a curve are defined as lines such that the distance between the curve and the line becomes arbitrarily small as either x or y tends to infinity. In the case of the function $y = 1/x$, the asymptotes are the x and y axes. We have already seen that as x tends to zero, y tends to either minus or plus infinity, depending on whether we take a left or right limit. In either case, the curve gets closer and closer to the y axis. Hence, the y axis provides one asymptote for this curve. Similarly, as x approaches either plus infinity or minus infinity, the value of $1/x$ approaches zero. Therefore, the x-axis also acts as one of the asymptotes of this curve. In this case, we say that $f(x)=1/x$ tends to zero asymptotically as x tends to infinity.

EXAMPLE

Consider a firm facing a demand curve of the form $p(q)=aq^{-b}$ where a and b are positive parameters. What are the properties of this demand curve?

Note that since this is a demand curve and q is the quantity of the good produced by the firm, we only need to consider nonnegative values of q. $p(0)$ is not defined but $p(q)$ is defined for all positive values of q. We can therefore set the domain of the function as the open interval $(0,\infty)$. The asymptotes of this function are $\lim_{q\to 0} p(q)=\infty$ and $\lim_{q\to\infty} p(q)=0$. Therefore, the asymptotes of this function are the vertical and horizontal axes of the Cartesian plane. Sketching the function for $a=1$ and $b=0.5$ gives the graph shown in Figure 2.8.

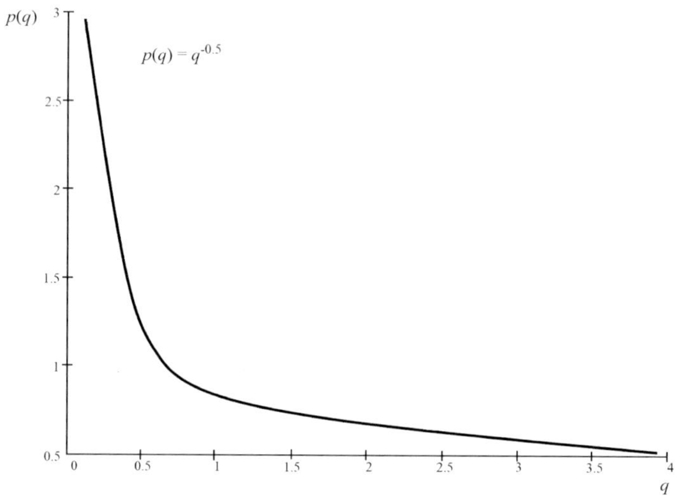

FIGURE 2.8 Plot of demand curve $p(q)=q^{-0.5}$.

It is often quite easy to evaluate the limits for simple functions, but more complicated functions can take a bit more work. However, there are some rules for combining simple limits which can make life a little bit easier. Suppose we have $\lim_{x\to c} f(x) = a$ and $\lim_{x\to c} g(x) = b$, where a and b are real numbers, then we can combine these limits using the following rules.

The sum-difference rule

The limit of the sum (or difference) of two functions is equal to the sum (or difference) of the limits.

$$\lim_{x\to c}\{f(x) \pm g(x)\} = \lim_{x\to c} f(x) \pm \lim_{x\to c} g(x) = a \pm b$$

EXAMPLE

Consider the equation $y = f(x) = 4x + 1/(x-2)$. What is the limit of y as x tends to the value 3?

We have $\lim_{x\to 3}\left(4x + \dfrac{1}{x-2}\right) = \lim_{x\to 3} 4x + \lim_{x\to 3} \dfrac{1}{x-2}$ by the sum rule. The first limit is simply equal to 12, and the second limit is equal to 1. By the sum rule, it follows that the limit of $f(x)$ as $x \to 3$ is equal to 13.

The product rule

The limit of the product of two functions is equal to the product of the limits.

$$\lim_{x\to c}\{f(x)g(x)\} = \lim_{x\to c} f(x) \lim_{x\to c} g(x) = ab$$

EXAMPLE

Let $y = f(x) = x\left(2 + \dfrac{1}{x}\right)$. What is the limit of $f(x)$ as x tends to 1?

We have $\lim_{x\to 1} x\left(2 + \dfrac{1}{x}\right) = \lim_{x\to 1} x \lim_{x\to 1}\left(2 + \dfrac{1}{x}\right)$ by the product rule. The first limit is equal to 1, and the second is equal to 3. Hence, $f(x) \to 3$ as $x \to 1$.

The quotient rule

The limit of the ratio of two functions is equal to the ratio of the limits, providing that the limit of the function which defines the denominator is not zero.

$$\lim_{x\to c}\left\{\frac{f(x)}{g(x)}\right\} = \frac{\lim_{x\to c} f(x)}{\lim_{x\to c} g(x)} = \frac{a}{b} \quad \text{if } b \neq 0.$$

EXAMPLE

Find the limit of $f(x)=\dfrac{3x^2}{1-1/x}$ as x tends to 2.

We have $\lim_{x\to 2} 3x^2 = 12$ and $\lim_{x\to 2}(1-1/x)=1/2$ which is not equal to zero.

Therefore, by the quotient rule, we have $\lim_{x\to 2}\left(\dfrac{3x^2}{1-1/x}\right)=\dfrac{12}{1/2}=24.$

The composition rule

If f is continuous at b, then the limit of the composition of the two functions f and g can be evaluated using the following relationship

$$\lim_{x\to c} f(g(x)) = f\left(\lim_{x\to c} g(x)\right) = f(b).$$

EXAMPLE

Find the limit of $f(x)=\left(\dfrac{x}{1+x^2}\right)^2$ as x tends to 1.

From the composition rule we have $\lim_{x\to 1}\left(\dfrac{x}{1+x^2}\right)^2=\left(\lim_{x\to 1}\left(\dfrac{x}{1+x^2}\right)\right)^2=\left(\dfrac{1}{2}\right)^2=\dfrac{1}{4}.$

It is important to note that we cannot apply these rules when the limits of either $f(x)$ or $g(x)$ are infinite. This is because the term "infinity" and the symbol ∞ do not refer to numbers in the conventional sense. If we do make the mistake of treating infinite limits as conventional limits, then we quickly run into paradoxical results.

REVIEW EXERCISES – SECTION 2.3

1. For each of the following functions, find the limit of $f(x)$ as $x \to c$.
   - **(a)** $f(x)=\dfrac{1}{\sqrt{x}}$ $c=\infty$
   - **(b)** $f(x)=(x^2+1/x)^3$ $c=2$
   - **(c)** $f(x)=4x^2+\dfrac{1}{x}$ $c=0$
2. Show that the function $f(x)=\dfrac{1}{x-1}$ is not defined at $x=1$ and derive its left and right limits as $x\to 1$. Use your answer to sketch the function.
3. For $f(x)=\dfrac{(2+x)^2-4}{x}$ evaluate the limit as $x\to 0$.

2.4 POWER FUNCTIONS

> Power functions are functions that take the form x^a, where x is a real number, and a is a fixed parameter. The linear function is an obvious example in which the input is simply raised to the power one. However, the power function is more general than this and can be used flexibly to produce very general shapes for the relationship between the input and output values.

Consider the function $y = f(x) = x^a$ where x is a real number. The symbol a represents the *parameter* of the function, that is, it is a number which is fixed for any given function but can be varied to create different functions. We will initially assume that a is one of the natural numbers, but this can easily be relaxed to cases in which a is a real number. If a is a natural number, then the range of the function depends on whether it is odd or even. If a is even, then the output of the function will always be positive irrespective of the sign of x. If a is odd, and x is negative, then the output of the function will also be negative.

Functions of this form are referred to as *power functions* since they are defined by raising the variable x to some power given by the parameter a. These functions are straightforward to manipulate, as we will now demonstrate. First, we note that the multiplication of two power functions can be achieved by adding the powers or *exponents* of the original functions. That is, we have:

$$x^a x^b = x^{a+b}. \tag{2.7}$$

This is easily demonstrated in the case where a and b are natural numbers with an example. Suppose we wish to multiply x^2 by x^3. We have $x^2 = x \times x$ and $x^3 = x \times x \times x$, and therefore $x^2 x^3 = x \times x \times x \times x \times x = x^5$. This generalizes to all cases in which a and b are natural numbers.

Another useful property is that raising a power function to some other power is achieved by multiplying the exponents. That is:

$$\left(x^a\right)^b = x^{ab}. \tag{2.8}$$

This can again be illustrated using an example. Suppose we wish to calculate $\left(x^3\right)^2$. This can be expressed as $\left(x \times x \times x\right)^2 = \left(x \times x \times x\right) \times \left(x \times x \times x\right) = x^6$. Again, this generalizes for any situation in which a and b are natural numbers.

Dividing one power function by another involves the subtraction of powers. Thus, we have

$$\frac{x^a}{x^b} = x^{a-b}. \tag{2.9}$$

For example, if we have x^3 / x^2, then this can be calculated as $x^{3-2} = x^1 = x$. This property allows us to demonstrate the important special case that $x^0 = 1$ since, if $b = a$, we have

$$1 = \frac{x^a}{x^a} = x^{a-a} = x^0.$$

This extends the set of numbers that the exponent can take from the set of natural numbers to include the number zero.

We can extend the set of values for the exponent to include the negative integers as follows. We have already established that when a and b are natural numbers, then $x^a / x^b = x^{a-b}$, for $x \neq 0$, and we also have $x^a x^b = x^{a+b}$. Therefore, dividing by x^b is equivalent to multiplying by x^{-b}, which means that the inverse of x^a can be written as x^{-a} and we have $x^a x^{-a} = x^0 = 1$. We can therefore write the equation $y = 1 / x^a$ as $y = x^{-a}$. This form of the equation is often neater and allows for the simplification of complicated expressions. Note, however, that we need to be careful in specifying the domain of the function. For example, for negative values of a, the input x cannot take on the value zero or the function will not be defined. As an example, consider the case $a = -1$, for $y = f(x) = x^a$, $a = -1$ gives us $y = 1 / x$, and we have already seen that this is not defined for $x = 0$.

We can also extend the definition of power functions to include cases in which the parameter is a rational number. This is useful because it allows us to express roots in terms of power functions. Consider, for example, the expression $x^{1/2}$. If we multiply this expression by itself then, by the rules for multiplication we have set out, we have $x^{1/2}x^{1/2} = x^1 = x$. We can therefore interpret $x^{1/2}$ as the square root of x. Similarly, $x^{1/3}$ gives us the cube root of x and, in general, the expression $x^{1/a}$, where a is a natural number, gives us the a'th root of x. We must again be careful in the specification of the function's domain when using expressions like this. For example, in the case of the relationship $f(x) = x^{1/2}$, we must restrict the domain to the nonnegative real numbers if the output of the function is to be real. However, this will not always be the case. For example, the relationship $f(x) = x^{1/3}$ determines the cube root of x, and this is defined for all real values of x, both positive and negative.

Finally, we can extend the class of power functions to include functions in which the exponent is a real number with a continuity argument. We have already extended the definition to include rational values of the parameter, and we know that any real number can be approximated to an arbitrary level of accuracy by a rational number. Therefore, we can also approximate the function $y = x^a$, where a is a real number, to any degree of accuracy we choose. Given this, we can apply power function rules when the exponent is any real number. This means that we can define, and manipulate, a very general group of functions under the general heading of power functions. Such functions are consistent with the general set of rules which we set in Table 2.2.

TABLE 2.2 Rules for power functions.

Multiplication	$x^a x^b = x^{a+b}$
Division	$\left(\frac{x^a}{x^b}\right) = x^a x^{-b} = x^{a-b};\ x \neq 0$
Powers	$(x^a)^b = x^{ab}$
Roots	$\sqrt[a]{x} = x^{1/a};\ \ a \neq 0$

where a and b are real numbers

REVIEW EXERCISES – SECTION 2.4

1. Simplify the following expressions using the rules for power functions.

(a) $f(x) = x^2 x^3$

(b) $f(x) = \dfrac{x^2}{\sqrt{x}}$

(c) $f(x) = x^a x^{-3}$

(d) $f(x) = (4x^2)^{-2}$

(e) $f(x) = \sqrt{4x^{-2}}$

2. For each of the following functions, we assume $0 < x < \infty$. In each case, demonstrate that the function satisfies the necessary conditions for the existence of an inverse and derive the equation for the inverse function.

(a) $y = f(x) = x^3 - 2$

(b) $y = f(x) = \dfrac{1}{\sqrt{x}}$

(c) $y = f(x) = 2x^2 + \dfrac{x}{2}$

2.5 EXPONENTIAL AND LOGARITHMIC FUNCTIONS

Exponential functions look superficially similar to power functions, but here, the input is the exponent of the function, and the parameter is the base. A related function is the logarithmic function which is the inverse of the exponential function.

Consider the function defined by

$$y = f(x) = c^x. \tag{2.10}$$

This looks very similar to the power function which we discussed in Section 2.4 but, in this case, x is the input variable and c is the parameter. If the domain is the set of real numbers and c is a positive real number (not equal to one), then this equation defines a function that maps the real numbers to the positive real numbers. For example, setting $c = 10$ generates the function of the form $y = f(x) = 10^x$, which is shown in Figure 2.9.

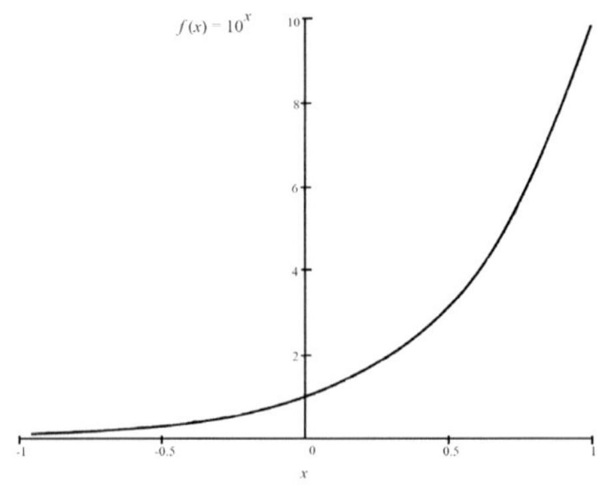

FIGURE 2.9 Graph of $y = 10^x$.

If the base is greater than one, then the exponential function is upward-sloping and has the same general shape as that shown in Figure 2.9. If the base is less than one, then the curve will be downward sloping but will still have the property that it maps the real numbers to the positive real numbers. For any value of the base, the curve will always cross the y-axis at the value one because $c^0 = 1$ for any value of $c \neq 0$.

Given that different values of the base produce essentially similar shaped functions, the choice of the base may seem unimportant. However, some bases are more convenient to work with than others. Base 10 is historically important because it was used to define the common logarithms used in calculation, but it is not the base that is most often used in mathematical analysis. Instead, mathematicians prefer to use the number e or *Euler's number* when working with exponential functions. This number can be derived as the sum of the infinite sequence shown in equation (2.11).

$$\begin{aligned} e &= \sum_{i=0}^{\infty} \frac{1}{i!} = 1 + 1 + \frac{1}{2} + \frac{1}{6} + \frac{1}{24} + \ldots \\ &= 2.781828\ldots \end{aligned} \tag{2.11}$$

The number e is a transcendental number. That is, it is a number that can be represented as an infinite nonrepeating decimal expansion which is not the root of a polynomial equation. This may seem to be a strange choice, but there are good reasons for its use which will become clear at a later stage. In fact, the number e is sufficiently important that mathematicians refer to it as the *natural base*, and unless there are particularly good reasons for choosing an alternative, it tends to be the default choice. This means that the function $y = f(x) = e^x$ is often referred to as *the* exponential function, despite the fact that e is just one of an infinite number of possible bases. The function is also frequently written as $y = f(x) = \exp(x)$. The exponential function with base e can be characterized in several different ways. One particularly useful way is as the power series shown in equation (2.12).

$$e^x = 1 + x + \frac{x^2}{2!} + \frac{x^3}{3!} + \ldots = \sum_{i=0}^{\infty} \frac{x^i}{i!}. \tag{2.12}$$

The properties of the exponential function are listed in Table 2.3. These properties hold for any choice of base c, where c is any positive real number that is not equal to one.

TABLE 2.3 Properties of the exponential function.

Multiplication	$c^{x_1} c^{x_2} = c^{x_1 + x_2}$
Division	$\frac{c^{x_1}}{c^{x_2}} = c^{x_1 - x_2}$
Powers	$(c^{x_1})^{x_2} = c^{x_1 x_2}$
Identity	$c^1 = c$
Zero exponent	$c^0 = 1$

We have already noted that if $c > 0$, then this function is always upward sloping. We, therefore, have a monotonic function which is defined for all real values of the input variable. It follows that an inverse function exists whose domain is the codomain of the original function. This inverse function is called the *logarithm* or *logarithmic (log)* function. The log function has domain equal to the set of positive real numbers and codomain equal to the complete set of real numbers. It can be written in the form shown in equation (2.13)

$$\begin{aligned} x &= f(y) = \log_c y \\ f &: \mathbb{R}_{>0} \to \mathbb{R} \end{aligned} . \tag{2.13}$$

The expression $\log_c y$ is read as "the log to the base c of y." Note that the log function is only defined for positive values of y, and is undefined for negative values, or for $y = 0$. When the natural base e is used, then we either write $x = \log_e y$ or $x = \ln y$. Figure 2.10 shows the log function for the natural base. Note that the log function will take this general shape for any base $c > 1$ and will always have the property that $x(1) = 0$.

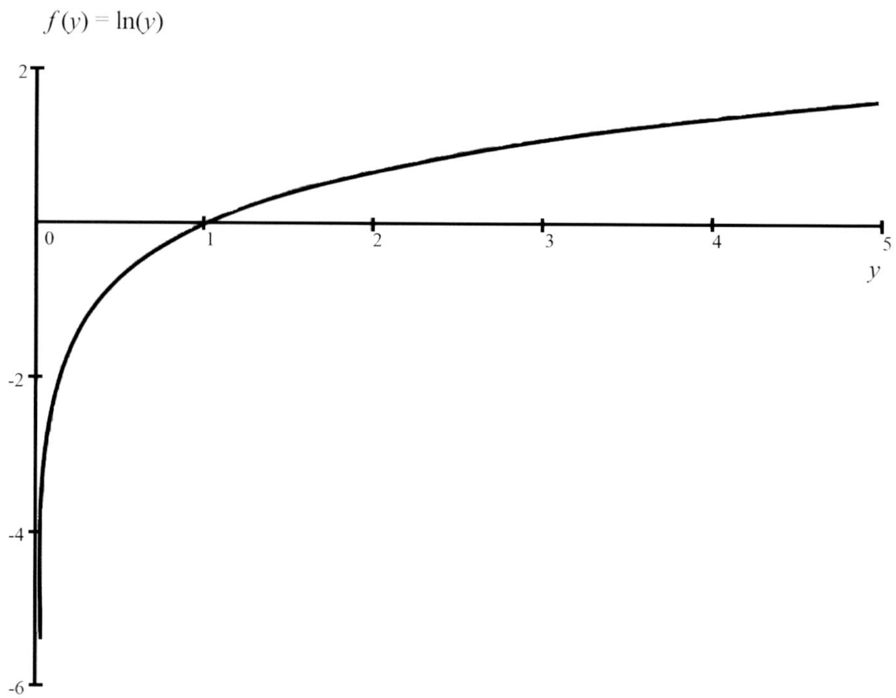

FIGURE 2.10 Graph of $x = \ln(y)$.

The properties of the log function follow directly from its definition as the inverse of the exponential function and are listed in Table 2.4. An important implication of these properties is that the log function can often be used to transform nonlinear relationships, involving products or ratios of variable, into linear relationships defined in terms of logarithms. This is an extremely useful property for many economic models because it is usually much easier to manipulate and solve models involving linear relationships.

TABLE 2.4 Properties of the log function.

Log of product	$\log_c(x_1x_2) = \log_c x_1 + \log_c x_2$
Log of ratio	$\log_c\left(\frac{x_1}{x_2}\right) = \log_c x_1 - \log_c x_2 \quad \text{for } x_2 \neq 0$
Log of power function	$\log_c x_1^b = b\log_c x_1$
Identity relationship	$\log_c c = 1$
Log of one	$\log_c 1 = 0$

Historically, the log function was extremely important in mathematics because it allowed the transformation of the relatively difficult operations of multiplication and division into the much simpler operations of addition and subtraction. For example, suppose we wish to divide one real number x_1 by another x_2. First, we take logarithms of the relevant numbers, $z_1 = \log_a x_1$ and $z_2 = \log_a x_2$. Next, we note that $\log_a(x_1 / x_2) = \log_a(y_1) - \log_a(y_2) = x_1 - x_2$. Finally, we reverse the operation of taking logs to obtain $(x_1 / x_2) = a^{z_1 - z_2}$. Multiplication can be carried out using a very similar procedure, except that we take the sum of the logarithms rather than the difference. This method was extremely useful before electronic calculators and computers became available, and most students would have books of "common logarithms" (logarithms to the base 10) and their corresponding "anti-logarithms," specifically for this purpose.

The use of logarithms for arithmetic calculation has declined since electronic calculators became the norm, but they remain important for many other reasons. A particular example here is the analysis of growth over time. Data shows that the aggregate output of many economies grows at a roughly constant proportional rate over long periods of time. This type of process can be captured by an exponential function of time, that is, an equation of the form $y(t) = y_0(1+g)^t$, where y_0 is the initial level of the variable and g is the annual growth rate. Thus, output is a nonlinear function of time events when the growth rate is constant. Figure 2.11 shows an index of UK Gross Domestic Product per capita for the period 1855 to 2019 with 1913=100. The slope of this graph appears to be getting steeper and steeper over time, but there is no acceleration in growth here. The increasing slope of the graph simply reflects the combination of a growing level of the variable with a constant proportional or percentage growth rate. Simple inspection of the graph of a growing variable can therefore give the misleading impression of accelerating growth.

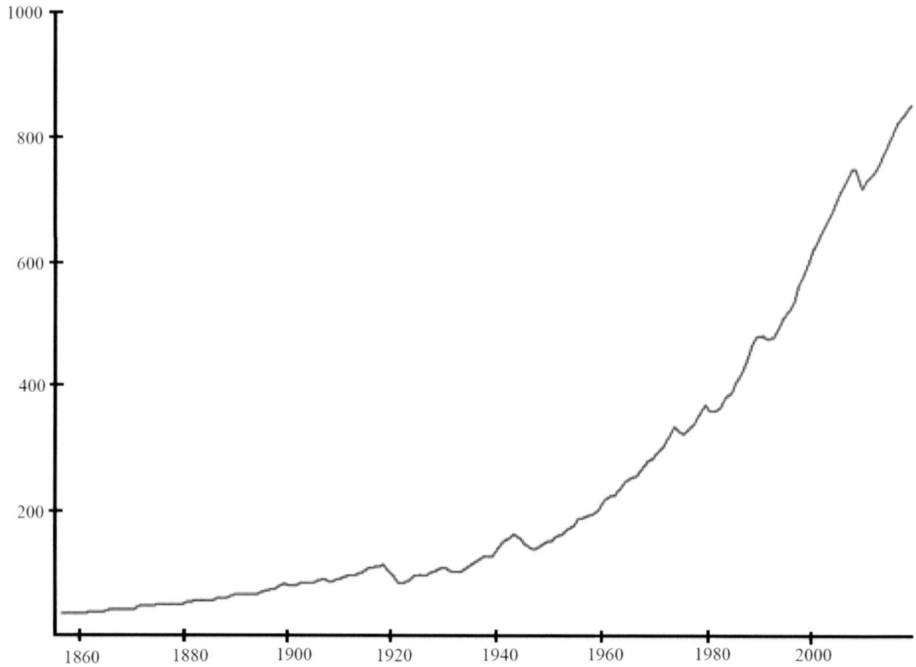

FIGURE 2.11 UK GDP per capita, 1913-100.

To better visualize the growth process, we take the logarithm of the series. If the series is growing at a constant proportional rate, then its value at time t is given by the exponential growth equation $y = y_0\left(1+g\right)^t$. Using the natural base e, and taking logarithms of this expression, yields $\log_e y_t = \log_e\left(y_0\right) + \log_e\left(1+g\right)\times t$, or $\ln y_t = \ln y_o + \ln\left(1+g\right)\times t$, where ln indicates the natural logarithm, or log to the base e. Setting $\ln\left(y_0\right) = \alpha$ and noting that, for small values of g, we can substitute $\ln\left(1+g\right) \simeq g$, this yields a linear relationship of the form $\ln y_t = \alpha + gt$. Thus, we obtain a relationship that is linear in logarithms. Figure 2.12 shows the natural logarithm of the series shown in Figure 2.11.

As expected, this shows an approximately linear relationship with time, thus indicating that a constant proportional growth rate is a reasonable approximation for this variable.

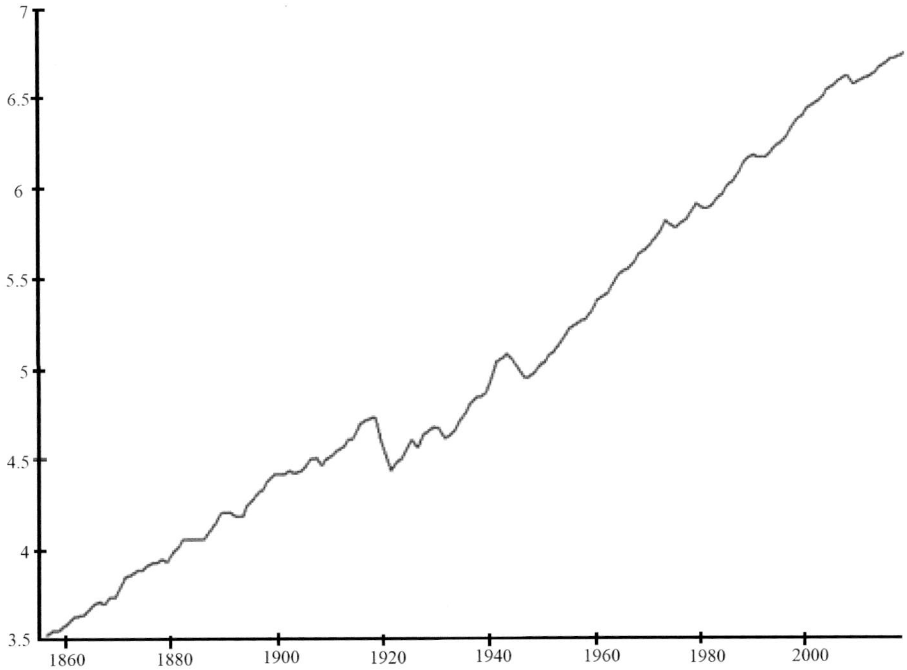

FIGURE 2.12 UK GDP per capita (log scale).

REVIEW EXERCISES – SECTION 2.5

1. Given the function $f(x) = 2^x$, where $-\infty < x < \infty$, evaluate the following, and use your answers to sketch the function.

(a) $f(2)$

(b) $f(1)$

(c) $f(0)$

(d) $f(-1)$

(e) $f(-2)$

2. Given $y = \log_2 x$, find the values of x which are consistent with
   (a) $y = 4$
   (b) $y = 1$
   (c) $y = 3$
3. Without using your calculator, show that $\ln(32) = 5\ln(2)$.

2.6 POLYNOMIAL FUNCTIONS

> The term "polynomial function" is used to describe a general class of functions that includes linear, quadratic, and cubic relationships, as well as terms with higher-order powers of the input variable. The use of higher-order powers in such functions allows for very general shapes.

A polynomial function is defined as a function that involves nonnegative integer powers of the variable x. The general form of such functions can be written as

$$\begin{aligned} f(x) &= a_n x^n + a_{n-1} x^{n-1} + \ldots + a_1 x + a_0 \\ &= \sum_{i=0}^{n} a_i x^i . \end{aligned} \tag{2.14}$$

where the input variable x is a real number. A function of the form (2.14) is referred to as an *nth order polynomial function* because n is the highest power of x included.

We have already seen that linear functions produce a straight-line relationship in the Cartesian plane. If we introduce higher-order powers into the relationship, then the shape of the output function will change. For example, a quadratic function takes the form $f(x) = a_2 x^2 + a_1 x + a_0$. When drawn in the Cartesian plane, this produces a curved relationship, the slope of which will change around some critical point. Consider, for example, the case shown in Figure 2.13, where we have $a_0 = 0$, $a_1 = 1$, and $a_2 = 1$. This produces the curve shown in the diagram, in which the slope is negative for values of x less than $-1/2$, and positive for values of x greater than $-1/2$. We also see that it cuts the x axis at two points, where $x = 0$ and $x = -1$. The introduction of cubic terms into a polynomial function will produce even more general shapes. For example, if we graph the function $f(x) = 2x^3 - 2x^2 + x$, as shown in Figure 2.14. We can observe that it has two turning points and cuts the x-axis in three places.

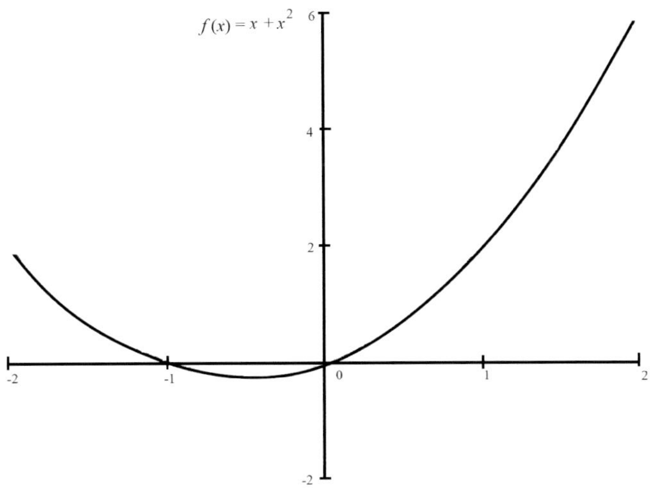

FIGURE 2.13 A quadratic function in the Cartesian plane.

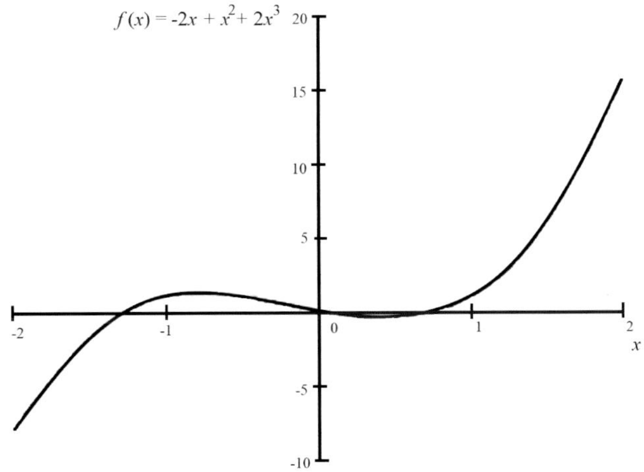

FIGURE 2.14 A cubic function in the Cartesian plane.

Turning points are defined as points at which the slope of the function changes sign, and the *roots*, or *zeros*, of the function are defined as points at which $f(x)=0$. If the roots are real, then the condition $f(x)=0$ means that the function cuts the x-axis at such points. As the order of the function increases, the potential number of turning points and real roots increase. However, this is not necessarily the case. For example, the fourth-order polynomial function

$f(x) = 1 + x^4$, has only a single turning point and does not cross the x-axis for any real value of x. The order of the polynomial puts an upper limit on both these features. For example, we can say that a cubic function has *at most* two turning points, and that it cuts the x-axis *at most* three times. In general, the maximum number of turning points is one less than the order of the polynomial (n-1), and the maximum number of real roots is equal to the order (n) of the polynomial.

EXAMPLE

Consider the polynomial function $f(x) = x^2 - 9x + 20$. This is a quadratic function so we can immediately tell that there are at most two real values of x which are consistent with $f(x) = 0$, and that there is at most one turning point. Factorizing the function to obtain $x^2 - 9x + 20 = (x-4)(x-5)$ tells us that the function crosses the x-axis at $x = 4$ and $x = 5$. It follows that the turning point of the function must occur somewhere between these values. We can confirm this with the plot of the function shown in Figure 2.15, which indicates a turning point at $x = 4.5$.

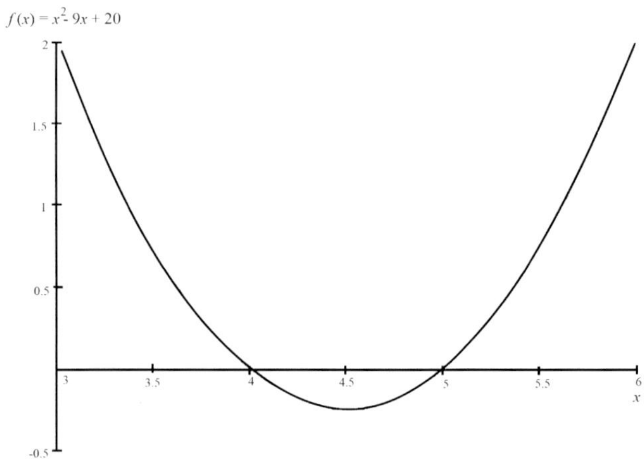

FIGURE 2.15 Plot of the polynomial function $x^2 - 9x + 20$.

If we allow x to take complex values, then it is possible to find roots for functions that do not cross the x-axis at any point. For example, consider the polynomial relationship $f(x) = x^2 + 2x + 5/4$. If we plot this relationship on the Cartesian plane, then we get the curve shown in Figure 2.16, which does not cross, or even touch, the x-axis. This means that there are no real values of x that are consistent with $f(x)=0$. However, we can show that the values $x_1=-1-i/2$ and $x_2=-1+i/2$ both give $f(x)=0$. Therefore, the roots of this polynomial function are complex conjugates.

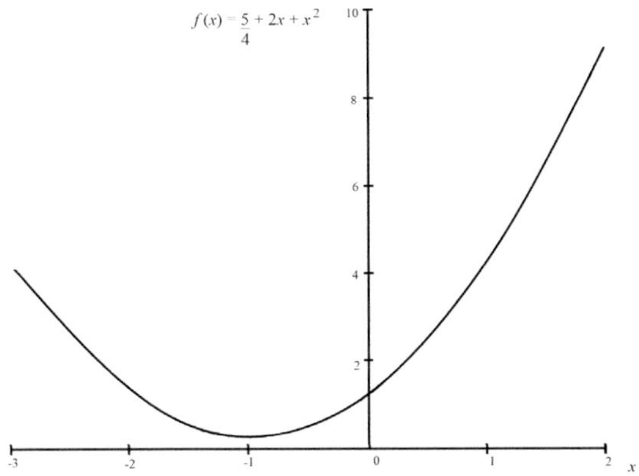

FIGURE 2.16 A quadratic function with complex roots.

In general, we can say that an nth order polynomial equation will have n roots, providing we allow x to take both complex and real values. In the case of real roots, some solutions may not be distinct. Table 2.5 gives some examples which clarify this point.

TABLE 2.5 Roots of quadratic functions.

Distinct real roots	$f(x)=x^2-1$	$x=1$ or -1
Repeated real roots	$f(x)=x^2-2x+1$	$x=1$
Complex roots	$f(x)=x^2+1$	$x=i$ or $-i$

Since any nth order polynomial has n roots (although some roots may be repeated and some may be complex), this means that it can be factorized and written in the form shown in (2.15)

$$\begin{aligned} f(x) &= b_0(x-b_1)(x-b_2)\ldots(x-b_n) \\ &= b_0\prod_{i=1}^{n}(x-b_i) \end{aligned} \tag{2.15}$$

where $b_i; i=1,\ldots,n$ are the roots. These roots contain important information about the nature of the function, and it will prove useful to find methods that allow us to solve for them. This is straightforward for low-order polynomials but becomes progressively more difficult as the order of the polynomial increases. For example, consider the linear function $f(x) = a_0 + a_1 x$. In this case, there is a single root, and it is trivial to solve for it by setting $f(x) = 0$, which yields $x = -a_0 / a_1$. In the case of quadratic equations, there are two roots, and we can solve for these either by factorization or by the general formula for quadratic equations that we introduced in Chapter 1. Although there are general formulas for finding the roots of cubic and quartic equations, these are much more difficult to apply than the quadratic formula and, once we have polynomials of order 5 or higher, there are no general formulas available to us.

In the case of cubic equations, it is sometimes possible to solve for the roots by guessing one of them and then factorizing the expression using this information. This reduces the problem finding the roots of a quadratic expression for which there is a standard solution.

EXAMPLE

Suppose we wish to solve for the roots of the cubic function $f(x) = x^3 + x^2 - 2x$. In this case, it is obvious that $x = 0$ is a root because we can see that $f(0) = 0$ by simple inspection of the function. Since the function factorizes to give $f(x) = x(x^2 + x - 2)$, we can solve for the remaining two roots by solving the quadratic equation $x^2 + x - 2 = 0$. This factorizes very easily to give $x^2 + x - 2 = (x-1)(x+2)$. Therefore $x = 1$ and $x = -2$ are also roots of this function. These solutions are confirmed by the plot of the function shown in Figure 2.17, which shows the function intersecting the x-axis at the three points we have identified.

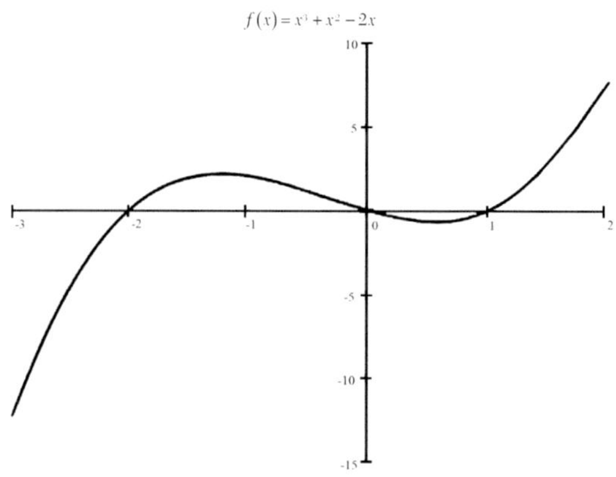

FIGURE 2.17 Roots of a cubic polynomial function.

It is not always easy to find the roots of higher-order polynomial functions analytically. However, we can often use numerical methods to find solutions when analytical methods fail. We have already seen an example of this in Chapter 1, Figure 1.6, which shows the *bracketing method* for finding roots. The bracketing method makes use of the *intermediate value theorem,* which we state below.

The Intermediate Value Theorem: If a continuous function has values of opposite signs at the endpoints of an interval, then the function has at least one zero (or root) within that interval.

The bracketing method works by starting with an interval that has the required property that its values have an opposite sign at the endpoints and then successively narrows that interval until the endpoints are sufficiently close to each other to constitute a solution. The most important prerequisite for this method to work is that we must be able to identify an initial interval when the function has values of opposite signs at the endpoints. If we can do this, then the bracketing method provides a robust method for finding a solution, although it can be inefficient in that it may require more calculations than some alternative methods.

EXAMPLE

Consider the function $f(x) = x^3 - 4.73x^2 - 3x + 14.16$. This is a cubic function, so we know that it has at most three distinct real roots, though there may be fewer. If we plot the function, as shown in Figure 2.18, then we see that, in this case, there are three distinct roots.

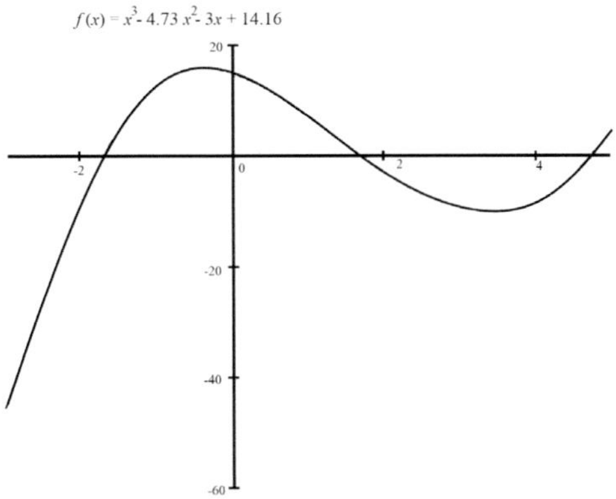

FIGURE 2.18 Plot of the function $f(x) = x^3 - 4.73x^2 - 3x + 14.16$.

We can use the information shown in Figure 2.18 to get more precise numerical solutions for the roots. First, we note that there is a root somewhere in the interval $[-2,0]$. Using the algorithm shown in Figure 1.6 and setting the limits of the interval at these values, we obtain the solution $x = -1.7307$. Next, we note that the interval $[0,2]$ also contains a root. Therefore, setting these as the endpoint values, we use the algorithm to obtain our second solution as $x = 1.7292$. Finally, we note that the interval $[2,5]$ contains a root, and that application of the algorithm in this case gives us $x = 4.7315$.

Note that the bracketing algorithm works best when we can identify intervals for which the output of the function changes sign at the endpoints. If this condition is not met, then it is not guaranteed that we will find a solution. However, failure of this condition does not mean that a solution does not exist. For example, suppose we chose an interval $[-2,2]$ for our function. The value of the function is negative at both endpoints even though there are two roots within this interval. Alternatively, suppose we chose the interval $[2,4]$. Again, the value of the function is negative at both endpoints but, in this case, there is no root in this interval.

REVIEW EXERCISES – SECTION 2.6

1. Find the roots of the following polynomial functions.

(a) $f(x) = x^2 - 5x + 6$

(b) $f(x) = x^2 - 6x + 9$

(c) $f(x) = 2x^2 + 3x + 1$

(d) $f(x) = 3x^2 + x - 2$

(e) $f(x) = x^2 - 4x + 5$

2. For the general quadratic function $f(x) = x^2 + bx + c$, show that

(a) If $b^2 > 4c$ then the function has distinct real roots.

(b) If $b^2 = 4c$ then the function has repeated real roots.

(c) If $b^2 < 4c$ then the roots are complex conjugates.

2.7 SINE, COSINE, AND TANGENT FUNCTIONS

The sine and cosine functions are based on trigonometric relationships. For a right-angled triangle, the sine is the ratio of the length of the opposite side to the hypotenuse. The cosine is the ratio of the length of the adjacent side to the hypotenuse. These relationships are illustrated in Figure 2.19.

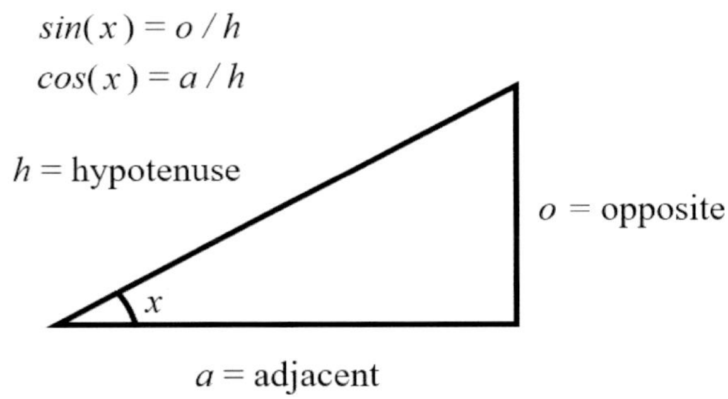

FIGURE 2.19 Sine and cosine functions.

For the angle x, we now define the sine function $y = \sin(x)$ and the cosine function $y = \cos(x)$ as illustrated in Figure 2.19. The domain of both these functions is the set of real numbers. Both the sine and the cosine functions are *cyclic,* meaning that as x increases, the output of the function repeats in the form of a cycle. The increase in x needed for the cycle to repeat depends on the units of measurement. For example, if the angle x is measured in radians, then the sine function goes through a complete cycle when x increases by 2π. The same is true for the cosine function.

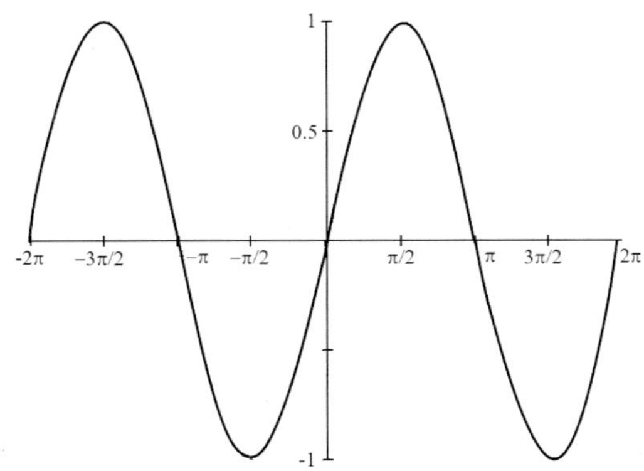

FIGURE 2.20 Plot of $y = \sin(x)$ for $-2\pi \le x \le 2\pi$.

Figure 2.20 illustrates the sine function through two complete cycles, as x increases from -2π to 2π. Note that the value of $\sin(0) = 0$ and the function reaches its maximum value of one when $x = \pi / 2$ and when $x = -3\pi / 2$. The minimum value of -1 is attained when $x = 3\pi / 2$ and when $x = -\pi / 2$.

If we define the sine function for a restricted domain that consists of one cycle, that is for $0 \le x < 2\pi$, then we can find an inverse function. This is written as either $x = \sin^{-1}(y)$ or $x = \arcsin(y)$. This gives the angle x which is consistent with a particular value of y. For example, we have $\sin^{-1}(\pi / 2) = \arcsin(\pi / 2) = 1$.

The cosine function has very similar properties to the sine function. Like the sine function, it is cyclic and goes through a complete cycle when x increases by 2π radians. It is also bounded by the values one and minus one like the sine function. Sine and cosine differ in that the values of the cosine function are offset from those of the sine function according to a fixed difference in the x values. For example, we have $\cos(0) = 1$ and $\cos(\pi / 2) = 0$. The cyclic nature of both these functions means that they are often used to model periodic or cyclical behavior in economic variables.

As with the sine function, we can define an inverse function for the cosine by defining it on a limited domain consisting of a single cycle, that is $0 \le x < 0$. The inverse of the cosine function is written as either $x = \cos^{-1}(y)$ or $x = \arccos(y)$. This gives the angle x, which is consistent with a particular value of the cosine function. For example, we can write $\cos^{-1}(0) = \arccos(0) = 1$.

Both the sine and the cosine functions can be represented as infinite series. For the sine function, we have

$$\sin(x) = x - \frac{x^3}{3!} + \frac{x^5}{5!} - \frac{x^7}{7!} + \ldots = \sum_{i=0}^{\infty} \frac{(-1)^i}{(2i+1)} x^{2i+1} \tag{2.16}$$

and for the cosine function, we have

$$\cos(x) = 1 - \frac{x^2}{2!} + \frac{x^4}{4!} - \frac{x^6}{6!} + \ldots = \sum_{i=1}^{\infty} \frac{(-1)^i}{(2i)!} x^{2i}. \tag{2.17}$$

These representations are useful in several different contexts. For example, when developing calculus, we can use these representations to demonstrate important results such as the fact that the cosine function is the derivative of the sine function.

Finally, we note that there are many trigonometric identities associated with the sine and cosine functions. For example, we can show that the sum of the squared values of the sine and cosine function for given values of x is equal to one, that is

$$\sin^2(x) + \cos^2(x) = 1.$$

Rather than attempting a thorough review of these identities at this stage, we will introduce them as and when necessary for particular applications.

Finally, we note that there is a third ratio of interest associated with the right-angled triangle shown in Figure 2.19, in the form of the *tangent,* which is written as $\tan(x)$. This is defined as the ratio of the length of the opposite side to the adjacent side, that is $\tan(x) = o / a$. Unlike the sine and cosine functions, this relationship is not cyclic. Moreover, it is not defined for all real values of x. For example, if $x = \pi / 2$, then the length of the adjacent side of the triangle is equal to zero, and therefore $\tan(\pi / 2)$ is not defined. By restricting the domain, we can however define a function of the form $y = \tan(x)$ where $-\pi / 2 < x < \pi / 2$. The graph of this function is illustrated in Figure 2.21. We see that the value of $\tan(x)$ tends to infinity as x tends to $\pi / 2$ from below, and to minus infinity as x tends to $-\pi / 2$ from above.

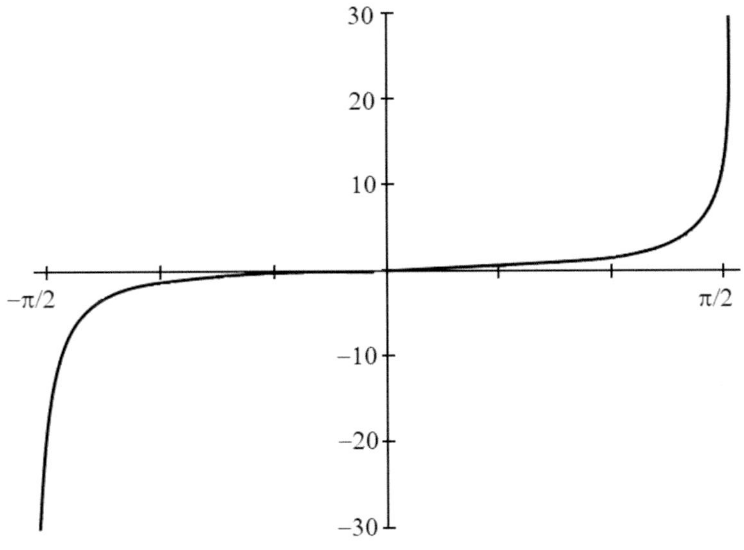

FIGURE 2.21 $y = \tan(x)$ for $-\pi / 2 < x < \pi / 2$.

Note that, as with the sine and cosine functions, we write the inverse of the tangent function as either $x = \tan^{-1}(y)$ or $x = \arctan(y)$. This gives us the value of the angle x which is consistent with a particular value of the function y.

For example, we have $\tan^{-1}(0) = \arctan(0) = 0$. Finally, we note that the sine, cosine, and tangent functions are linked by the identity $\tan(x) = \sin(x) / \cos(x)$.

REVIEW EXERCISES – SECTION 2.7

1. Find the following for a right-angled triangle with opposite side equal to 1 and adjacent side also equal to 1.
   - **(a)** The angle x
   - **(b)** $\tan(x)$
   - **(c)** $\sin(x)$
   - **(d)** $\cos(x)$
2. Show that the equation $\sin^2(x) + \cos^2(x) = 1$ is true for any angle x.
3. Let x be an angle that is measured in radians and $0 \le x < 2\pi$. Plot the function $y = f(x) = \sin(2x)$.

CHAPTER 3

SIMULTANEOUS EQUATIONS

Economic and business analysis frequently requires us to seek the solution of systems of simultaneous equations. For example, the analysis of markets involves the solution of demand and supply systems for the equilibrium price and quantity values. In macroeconomic analysis, the Keynesian model of output determination is written as a simultaneous system of equations in output, consumption, and autonomous expenditures, which we solve to find an equilibrium. This chapter explores the mathematics of systems like the Keynesian model. Our aim is to determine the conditions necessary for a solution to exist and to find methods through which we can systematically find the solution.

3.1 LINEAR EQUATIONS

Systems of linear equations are relatively simple to solve. In this section, we look at the properties of linear equations and show how they can be transformed into forms which make finding solutions easy.

A *linear equation* is a first-order polynomial function. The general form of such a relationship is given in equation,

$$y = a + bx \tag{3.1}$$

where y and x are variables which we will assume are real numbers. The symbols a and b represent parameters. That is, they are general symbols for numbers which are fixed for any given equation but can be varied for the purposes of analyzing different equations. The parameter a is the *intercept*, that is, the value of y at which the graph of the function crosses the vertical axis when the line is drawn in the Cartesian place. The parameter b is the *slope* or *gradient* of the line. This gives this ratio of the change in y divided by the change in x for a given interval on the line. The gradient of a linear equation is constant for any interval. This form of the equation is known as the *explicit form* because the *dependent variable*, y, is written explicitly in terms of the *independent variable*, x. A linear equation can be interpreted as a function which maps the set of real numbers to itself. This is true because the relationship is defined for every value of x in the set of real numbers, and, providing $b \neq 0$, the output of the equation will also consist of the entire set of real numbers.

An example of a linear equation is shown in Figure 3.1. The equation shown takes the form $y = 1 + 0.5x$. Thus, the intercept, or value of y when x=0, is given by 1 and the gradient $\Delta y / \Delta x$ is 0.5, where the symbol Δ or *delta* is used to indicate a change in either variable between two points. On the diagram, the gradient is calculated using the interval $x = 1$ to $x = 2$, which results in an increase in the value of y from $y = 1.5$ to $y = 2$, which therefore gives us $\Delta y / \Delta x = (2 - 1.5) / (2 - 1) = 0.5$. For linear equations, the gradient will be the same for any chosen interval. This graph can be extended indefinitely for any value of x in the interval $-\infty$ to ∞, and it is also the case that for any real number $y = y_1$ there is some value of $x = x_1$ such that $y_1 = a + bx_1$. Therefore, both the domain and the range consist of the full set of real numbers.

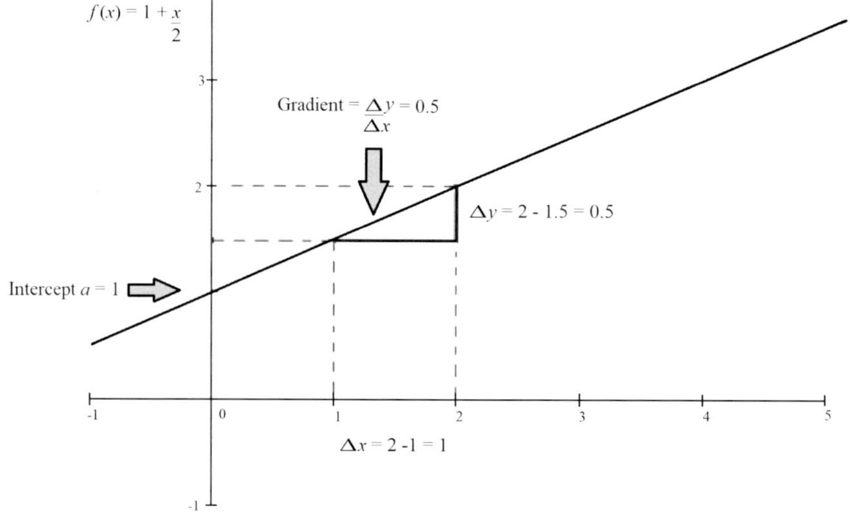

FIGURE 3.1 Parameters of a linear equation.

It is often useful to transform a linear equation to express it in a more convenient form. The following operations will allow us to do this.

(1) Addition or subtraction of a constant to both sides of the equation.

We can add or subtract a constant from both sides of the equation while maintaining the equality. For example, if $y = a + bx$, then $y + c = c + a + bx$ remains true for all real numbers c. If c is negative, then this is equivalent to subtracting a number from both sides. This rule also applies if we add or subtract terms which depend on the variables. For example, the equations $y = a + bx$ and $y - bx = a$ are equivalent. This property is useful if we wish to make x the subject or output of the equation.

EXAMPLE

Let $y = 4 + x$, subtracting 4 from both sides of the equation gives $x = y - 4$.

(2) Multiplication by a constant.

We can multiply both sides of an equation by a constant while maintaining the equality. Therefore, if $y = a + bx$, then $cy = ac + bcx$ remains true for all real numbers c. This property is useful if we wish to write an equation so that all its parameters are whole numbers.

EXAMPLE

Let $y = 1/3 + 2x$, multiplying through by 3 gives us an equation of the form $3y = 1 + 6x$.

(3) Division by a constant.

If we divide both left and right-hand sides by a nonzero constant, then the equation remains valid. Therefore, if $y = a + bx$, then $y/c = a/c + bx/c$ remains true for all real numbers $c \neq 0$. This property is useful if we wish to write the equation so that the parameter associated with one of the variables is equal to one.

EXAMPLE

Let $20y = 60 + 40x$, dividing both sides by 20 gives us $y = 3 + 2x$.

Note that this property specifically excludes the number zero because division by zero is not a valid mathematical operation.

(4) Raising both sides to the same power.

If we raise both left and right-hand sides to the same power, then the equation remains valid. Therefore, if $y = a + bx$, then $y^c = (a + bx)^c$ remains true for all real numbers. This property is useful when working with nonlinear equations.

EXAMPLE

Let $y = 3x + 2$, squaring both sides of the equation gives us an equation of the form $y^2 = (3x + 2)^2 = 9x^2 + 12x + 4$.

These properties are useful when we wish to transform an equation and write it in an alternative format. So far, we have written equations in explicit form, that is we have made y the dependent variable of the equation and x the independent variable. Sometimes, however, it is more convenient to write equations in *implicit form* in which there is no distinction between dependent and independent variables. This is quite common in economics when the equation represents an equilibrium relationship between two variables rather than a causal relationship in which one variable determines the other. Implicit equations are usually written with all the variables on one side of the equation, for example, we might have $ax + by = c$, where x and y are variables; and a, b, and c are parameters.

Consider the relationship $y = 1 + 0.5x$ which is shown in Figure 3.1. To write this in implicit form, we multiply through by two to obtain $2y = 2 + x$, and then subtract x from both sides, to obtain the implicit form $2y - x = 2$. The implicit form of the equation is not unique because we can always multiply both sides by any real number to obtain an alternative representation. For example, multiplying our equation by two gives us $4y - 2x = 4$, which is an equally valid form of the same equation.

In the case of linear equations, we can use these rules to obtain the *inverse relationship*, providing $b \neq 0$. Consider the general case $y = a + bx$, subtracting a from both sides gives us $y - a = bx$, and then dividing both sides by b, gives us $x = -a / b + (1 / b)y$. The equation now has x as the subject, or dependent variable, and y as the input, or independent variable. For our example $y = 1 + 0.5x$, application of these steps gives us the inverse equation $x = -2 + 2y$. Note that all three forms of the equation that we have derived, that is $y = 1 + 0.5x$, $2y - x = 2$, and $x = -2 + 2y$, produce exactly the same line when graphed in the Cartesian plane. These are simply different ways of writing the same relationship in equation form, rather than different relationships.

REVIEW EXERCISES – SECTION 3.1

1. Find the equations of the straight lines which pass through the following pairs of points in the Cartesian plane.

(a) $(-1,1)$ and $(4,3)$

(b) $(2,5)$ and $(1,7)$

(c) $(1,4)$ and $(2,7)$

(d) $(-1,5)$ and $(4,5)$

2. For each of the following equations, which are written in explicit form, find the values of b and c which give an equivalent representation as implicit equations.

(a) $y=-2+x/3;\quad x+by=c$

(b) $y=4-5x;\quad bx+y=c$

(c) $y=3-6x;\quad bx+cy=6$

(d) $x=2-3y;\quad 2x+by=c$

3. Transform each of the following equations so that they take the form $x=b+cy$.

(a) $y=5+3x$

(b) $y=-3-2x$

(c) $y=10-4x$

(d) $4x+3y=2$

3.2 SYSTEMS OF LINEAR SIMULTANEOUS EQUATIONS

> In this section, we look at the process of solving pairs of linear simultaneous equations. This is relatively easy because the linear nature of the system limits the number of possible solutions. The methods we describe in this section can also be applied to nonlinear equations and to systems of many variables.

Suppose we wish to solve a pair of equations. Here, a solution means finding a pair of values for the unknown variables which are consistent with both equations. For example, suppose we have the equations given in.

$$\begin{aligned} ax + by &= c \\ dx + ey &= f. \end{aligned} \tag{3.2}$$

In this system x and y are variables, and a, b, c, d, e, and f are parameters. A solution is a pair of values x and y which is consistent with a particular set of parameter values. We can show that there will be a unique solution providing the two lines defined in are not parallel (i.e. have identical slopes), that is providing $a/b \neq d/e$.

To illustrate the process of finding a solution, we will begin with an example with specific parameter values. Suppose we have

$$\begin{aligned} 3x - 2y &= -2 \\ x + y &= 6. \end{aligned} \tag{3.3}$$

One method for finding a solution is to plot the equations and look for points of intersection. Applying this method to gives the graph shown in Figure 3.2. To find values of x and y which solve the system, we look for the point at which the two lines cross. In this case, it is easy to identify the solution as the point $(2,4)$ in the Cartesian plane.

The graphical solution of simultaneous equations is a good way of illustrating the existence of a solution but it is not very practical solution method for more complicated systems. Even for simple systems like, it can be time consuming and will usually involve some degree of error. Graphical methods can therefore sometimes be used to identify approximate solutions, but, in general, we will need to use numerical methods to find an accurate solution.

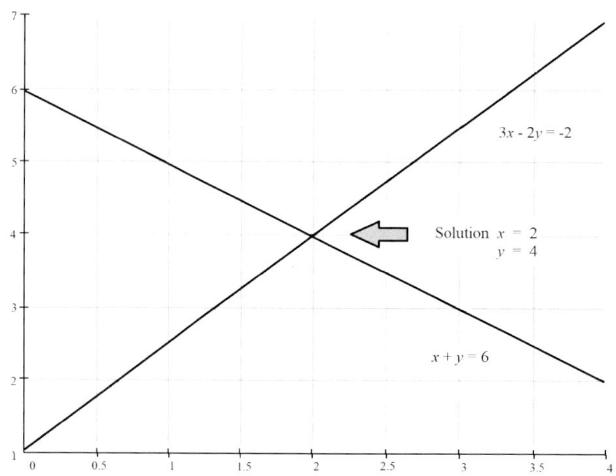

FIGURE 3.2 Simultaneous linear equations.

The first numerical method we will introduce is the *method of substitution*. This uses transformations of the equations in the system to obtain an equation which contains only one unknown variable. Once we have solved this equation, we can substitute the solution back into the other equation(s) of the system to solve for the remaining unknown variable(s). For example, consider the second equation in the system. This can be written in explicit form as $y = 6 - x$. Substituting this into the first equation gives the following expression

$$3x - 2(6 - x) = -2 \Rightarrow 5x = 10.$$

That is, we have reduced the system to a single equation in the single unknown variable x. It is easy to solve this equation to obtain $x = 2$ and we can then substitute this into either of the two original equations given in to obtain the solution for y. Substituting $x = 2$ into the second equation gives $2 + y = 6$, which gives the solution $y = 4$. This confirms the result we obtained earlier using the graphical method.

The method of substitution is probably the easiest numerical method to apply to pairs of simultaneous equations. In larger systems of equations, however, it becomes more difficult and other methods become more efficient. The most common method in larger systems is the *method of elimination* or *Gaussian elimination*. This is a systematic method, or algorithm, which can be applied in large systems of equations. It also has the advantage that it can easily be programmed for computer applications. Gaussian elimination takes linear combinations of the equations in the system to create a system which is easy to solve. Linear combinations are transformations of the system in which we either transform individual equations or add equations to each other in ways which change the presentation of the system but maintain the same equilibrium solution. The objective of these transformations is to represent the system in *triangular form*. This means we have a system in which one of the equations contains only one variable, the next contains that variable plus one other, and so on. Once the system is written in this form, the solution becomes very easy.

Let us consider how we can transform the system of equations given in into triangular form. If we multiply the second equation by two, then the system becomes

$$\begin{aligned} 3x - 2y &= -2 \\ 2x + 2y &= 12. \end{aligned}$$

Next, we add the first equation to the second equation, to write the system as

$$\begin{aligned} 3x - 2y &= -2 \\ 5x &= 10. \end{aligned}$$

The system is now in triangular form, with the first equation containing two variables, while the second contains only one. The second equation solves easily to give $x = 2$, and substituting this into the first equation, we obtain $y = 4$.

In this simple example, there is little to choose between the alternative methods we have described but, as we add more variables to the system, the advantages of Gaussian elimination become more obvious. In large systems, the systematic nature of the algorithm lends itself to implementation using computers. Therefore, this approach is the method used to solve simultaneous equation in most computer software. We will return to this method in a later chapter when we introduce matrix methods.

The solution methods we have described assume that a solution exists. This will not always be the case, even for linear systems. Before we start the process of looking for a solution, it is usually important to establish whether there is one to be found. In the case of linear equations, there are three possible outcomes. First, there may be a unique equilibrium solution of the kind we have assumed so far. Second, there may be no solutions. Third, there may be an infinite number of solutions. We can illustrate these possibilities for the general two-equation linear system defined in using the graphs shown in Figure 3.3.

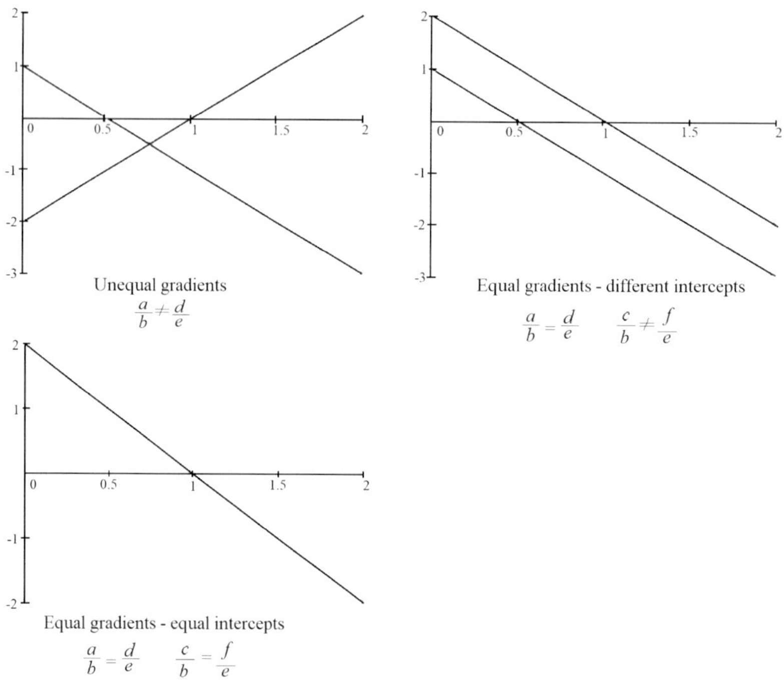

FIGURE 3.3 Possible cases for simultaneous linear equations.

(1) In the first case, there is a unique solution. This occurs if the lines defined by the equations in have different gradients and therefore intersect at a single point.

(2) In the second case, there are no solutions. This occurs if the lines have the same gradient but different intercepts. In this case, the equations define parallel lines which never intersect.

(3) Finally, in the third case, there are an infinite number of solutions. This occurs if the lines have the same gradient *and* the same intercept. In this case, the two equations define identical lines. This may not be immediately obvious if the equations are written in different ways.

A unique solution exists if, and only if, the gradients of the two lines are different. In the system, the gradient of the first equation is $-a/b$ and that of the second equation is $-d/e$. It follows that the condition for the existence of a unique solution in the system defined by can be written as $(a/b) \neq (d/e)$ or, alternatively, $ae \neq bd$. This gives us a condition for the existence of a solution which we can check before attempting to solve the system. We can derive a similar condition for systems of more than two linear equations, but this will require the use of matrix methods and will be covered in a later chapter.

REVIEW EXERCISES – SECTION 3.2

1. Graph the pair of simultaneous equations given below and use your graph to find an approximate solution.

$$y - x = 0 \qquad 4y + x = 5$$

2. Establish if the following systems of equations have a unique solution, no solution, or an infinite number of solutions.

(a) $2x - y = 0 \qquad -3x + 2y = 1$

(b) $4x + 2y = 1 \qquad x + y/2 = 2$

(c) $4x + y = 12 \qquad -3x + 2y = 2$

(d) $x - y/2 = 1 \qquad 2x - y = 2$

3. For the following pairs of simultaneous equations, establish that a unique solution exists and then find that solution using the method of substitution.

(a) $3x + y = 5 \qquad x - 2y = -3$

(b) $x - y/2 = 0 \qquad 2x + y = 4$

(c) $x+y=7$ $\quad 2x-y=5$

(d) $4x+y=13$ $\quad x-y=-3$

4. For the following pairs of simultaneous equations, establish that a unique solution exists and then find that solution using the method of elimination.

(a) $x+2y=7$ $\quad 3x-2y=5$

(b) $2x+y=2$ $\quad 4x+y=3$

(c) $4x+y=4$ $\quad x-y=1$

(d) $x+3y=3$ $\quad 2x-9y=1$

3.3 SOME EXAMPLES FROM ECONOMICS

> There are many examples of economic models which can be written in the form of linear simultaneous equations. In this section, we will look at two examples and show how these models can be solved using the methods discussed in Section 3.2.

Let us begin with a model which is covered in every introductory economics module: the two-equation model of demand and supply. This is one of the most basic models in economic analysis and is usually taught as part of an introductory course in microeconomics. For example, consider the pair of equations defined in object 3.4.

$$\begin{aligned} p &= 5-\frac{1}{2}q \qquad (1) \\ q &= 1+p \qquad (2) \end{aligned} \qquad (3.4)$$

Here, p is price and q is quantity and p and q are the *endogenous variables* of the system. Endogenous variables are variables which are determined within the system. In this case, p and q are determined by the interaction of demand and supply factors. The parameters of the system are the intercepts and slopes of the two curves.

The demand curve (1) is a downward sloping relationship in (q,p) space. Note that it does not really matter whether we make p or q the subject of the equation since both are endogenous variables. In practice, the choice of how we present this equation will depend on assumptions we make about the nature of the market we are describing. Here, p is on the left-hand side of the equation, and we refer to this as an *inverse demand curve*. For the purpose of solving the system however, there is no reason why the demand curve could not be written in the form $q = 10 - 2p$, since this would make no difference to the outcome. The supply curve (2) takes the form $q = 1 + p$ but could equally be written as $p = q - 1$ without changing the solution.

The easiest way to solve this system is by the method of substitution. Substituting equation (2) into equation (1) gives us an equation in one unknown variable p which can be solved easily for the market clearing price as shown in the following steps.

$$\begin{aligned} & p = 5 - 0.5(1 + p) \\ & \Rightarrow 1.5p = 4.5 \\ & \Rightarrow p = 3 . \end{aligned}$$

We can now substitute this into either the demand curve or the supply curve to determine the market clearing quantity. Using the supply curve, we have $q = 1 + p = 4$.

The method of substitution is easy to apply in small systems of equations in which some of the equations are set out in explicit form. This is true because it is straightforward in small systems to reduce the number of variables by substituting one equation into another. As the number of equations increases, however, this becomes increasingly difficult, especially when the equations of the model are not written explicitly. For larger systems, the method of Gaussian elimination can often provide a more efficient method of solution.

Let us consider an example of the Gaussian elimination method in practice. Consider the three-equation system set out in 3.5. This system describes a simple Keynesian income-expenditure model in which output Y, consumption expenditures C, and tax receipts T are jointly determined:

$$\begin{aligned} & Y = C + I + G && (1) \\ & C = 20 + 0.8(Y - T) && (2) \\ & T = 10 + 0.2Y && (3) \end{aligned} \qquad (3.5)$$

In addition to the three endogenous variables Y, C, and T, there are two exogenous variables, investment I and government spending G. The exogenous variables are assumed to be determined outside the system. The relationships between the variables of the model are defined by the model parameters, which are fixed numerical values.

The equations given in system reflect assumptions about the way in which the economy as a whole, the *macroeconomy*, works. Equation (1) is the national income accounting identity. It states that aggregate output Y is equal to the sum of consumption expenditures (C), investment expenditure (I), and government expenditure (G). Equation (2) is the consumption function. This reflects the assumption that aggregate consumption is a linear function of disposable income. The parameter 0.8 is the marginal propensity to consume, or the change in consumption in response to a unit change in disposable income. Equation (3) describes a simple model of the tax system and assumes that total taxation receipts are equal to the sum of an autonomous element (equal to 10), and an induced component $0.2Y$, where 0.2 is the marginal tax rate.

The system can be solved using the method of Gaussian elimination for given values of the exogenous variables. Let us assume that $G = I = 50$. Substituting these values into the system and rearranging so that the endogenous variables are on the left-hand side of the equations allows us to write the system as

$$\begin{aligned} Y - C \qquad\quad &= 100 \quad (1) \\ -0.8Y + C + 0.8T &= 20 \quad (2) \\ -0.2Y + \qquad\quad T &= 10 \quad (3) \end{aligned}$$

Next, we perform linear operations on these equations so that we can write the system in triangular form. The system will be in triangular form when equation (3) contains only one endogenous variable, equation (2) contains two endogenous variables, and equation (1) contains all three endogenous variables. Once the system is in triangular form, it will be easy to solve by the method of *backward substitution*. That is, we will first solve the third equation, then we will use the solution to solve the second equation and, finally, we will use both these solutions to solve the first equation.

To solve our system, we first multiply equation (1) by 0.8 and add the transformed equation to equation (2) so that the system becomes

$$\begin{aligned} Y - C \qquad\quad &= 100 \quad (1) \\ 0.2C + 0.8T &= 100 \quad (2) \\ -0.2Y + \qquad\quad T &= 10 \quad (3) \end{aligned}$$

Next, we multiply equation (1) by 0.2 and add the transformed equation to equation (3), to obtain the following

$$\begin{aligned} Y - C \qquad\quad &= 100 \quad (1) \\ 0.2C + 0.8T &= 100 \quad (2) \\ -0.2C + \quad T &= 30 \quad (3) \end{aligned}$$

Finally, we add equation (2) to equation (3) to obtain

$$\begin{aligned} Y - C \qquad\quad &= 100 \quad (1) \\ 0.2C + 0.8T &= 100 \quad (2) \\ 1.8T &= 130 \quad (3) \end{aligned}$$

The system is now in triangular form and can be solved easily by the method of backward substitution. First, we solve equation (3) for T to obtain $T = 130 / 1.8 = 72.22$. Substituting this into equation (2) then gives us $0.2C + 0.8 \times 72.22 = 100$, which solves to give $C = 211.12$. Finally, we substitute the solution for C into equation (1) to obtain $Y = 311.12$.

Although it may be easier to solve small systems using less formal methods, the advantage of the Gaussian elimination method is that it provides a systematic way of approaching the solution of systems of simultaneous equations. In particular, it lends itself naturally to problems which can be defined in matrix terms and can be easily implemented using numerical computing methods. This means that we can easily solve systems involving quite large numbers of variables. How do we know if a system of linear equations has a solution? For a system of linear equations to have a unique solution, we need the equations of the system to be linearly independent. Linear independence means that, if we choose any equation in the system, then it is not possible to find a linear combination of the other equations which is equal to our equation of choice. When we have a pair of linear equations, linear independence simply requires that the gradients of the two equations must not be equal. However, this becomes harder to establish in systems with three or more endogenous variables.

REVIEW EXERCISES – SECTION 3.3

1. Using the method of substitution, solve the following pairs of demand and supply equations.

   (a) $p = 102 - 2q$
   $q = 48 + p$

(b) $$\begin{aligned} p &= 19 - 0.75q \\ q &= 18 + 0.5p \end{aligned}$$

(c) $$\begin{aligned} p &= 14.5 - 0.25q \\ q &= 24.4 + 0.8p \end{aligned}$$

2. The following equations describe a Keynesian model of the open economy where the endogenous variables are national income Y, consumption C, and imports M. All other variables are exogenous.

$$\begin{aligned} Y &= C + I + G + X - M \\ C &= 30 + 0.7Y \\ M &= 10 + 0.4Y\ . \end{aligned}$$

Using the method of Gaussian elimination, solve for the equilibrium values of the endogenous variables when the values of the exogenous variables are $I = 100$, $G = 100$, and $X = 150$, where I, G, and X are investment, government spending, and exports.

3.4 NONLINEAR SIMULTANEOUS EQUATIONS

> Nonlinear systems of equations are often harder to solve than linear systems. There may be multiple solutions to nonlinear systems because the curves defined by the equations may intersect more than once.

If our system includes nonlinear equations, solving the system becomes more complicated because it is possible for more than one solution to exist. In fact, we will see that the number of solutions is much harder to establish by simple inspection of the system in such cases. However, it is often possible to determine the maximum number of solutions by identifying the *order* of the system.

Let us consider an example of a nonlinear system as shown in (3.6)

$$\begin{aligned} y &= 2x^2 \qquad (1) \\ y &= -4 + 6x \qquad (2) \end{aligned} \ . \tag{3.6}$$

This system will have two distinct solutions. We can show this easily by plotting the two curves defined in as shown in Figure 3.4. This shows the line representing equation (2) cutting the curve representing equation (1) in two places. In this case, we should therefore expect to find two distinct solutions when we solve the system numerically.

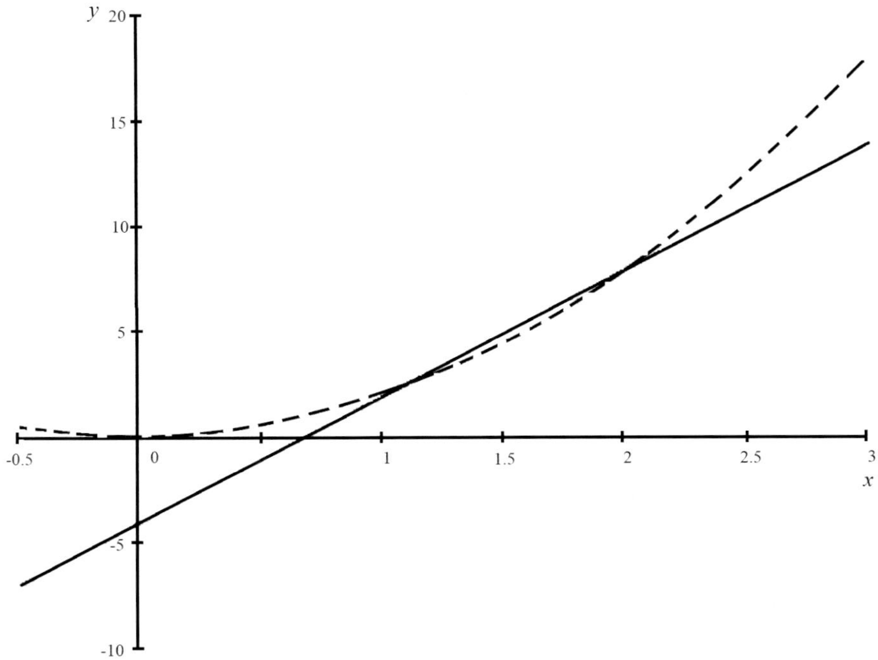

FIGURE 3.4 Solutions for a quadratic system of equations.

In this case, it is easy to solve the system using the method of substitution. We can eliminate y easily by subtracting equation (2) from equation (1) to obtain an equation of the form $2x^2 - 6x + 4 = 0$. This is a quadratic equation with a single unknown variable x and, therefore, has at most two real solutions. The equation factorizes easily to yield $2x^2 - 6x + 4 = 2(x-2)(x-1)$ and, therefore, the solutions for x are $x = 2$ and $x = 1$. We can obtain the corresponding solutions for y using either of the two original equations. This gives us two possible solutions for the system as either $(x,y) = (2,8)$ or $(x,y) = (1,2)$.

The number of distinct real solutions depends on the parameters of the system. For example, by changing the intercept of equation (2), we will change the number of real solutions. Suppose equation (1) remains the same, but we subtract 1/2 from the intercept of equation (2), which now becomes $y = -9/2 + 6x$. Applying the same procedure as before gives us a single equation of the form $2x^2 - 6x + 9/2 = 0$. This factorizes to yield $2x^2 - 6x + 9/2 = 2(x - 3/2)^2$. Therefore, there is single repeated root given by $x = 3/2$, and the system has a single solution equal to $(x,y) = (3/2, 9/2)$. If we were to draw the graph of the new system, we would see that the line defined by equation (2) is tangent to the curve defined by equation (1) at this point.

Next, suppose we subtract 1 from the original equation (2) so that it becomes $y = -5 + 6x$. In this case, the system can be reduced to a single quadratic equation of the form $2x^2 - 6x + 5 = 0$. This factorizes to yield complex roots of the form $x = 3/2 \pm i$. In this case, there are no real roots and no real solutions to the system of equations. If we were to plot the equations of this system, we would find that the curve $y = 2x^2$ always lies above the line $y = -5 + 6x$. Thus, when our system of equations consists of one quadratic equation and one linear equation, it is possible for it to generate two, one, or no real solutions depending on the parameters of the system.

In general, if we have a system of equations defined by polynomial relationships, then the higher the order of the polynomials in the system, the more real solutions are possible. Consider the system defined by, in which equation (1) is a cubic expression while equation (2) is linear

$$\begin{aligned} y &= x^3 + 4x^2 \qquad (1) \\ y &= 6 - x \qquad (2). \end{aligned} \tag{3.7}$$

This system has three distinct real solutions, as we can see from Figure 3.5, which shows the line (2) crossing the cubic function (1) in three places. To solve the system numerically, we will need to solve a cubic polynomial equation. Using the method of elimination, we can write the system as a single cubic equation of the form $x^3 + 4x^2 + x - 6 = 0$. This factorizes to yield $x^3 + 4x^2 + x - 6 = (x-1)(x+2)(x+3)$. We, therefore, have three solutions for x given by $x = 1$, $x = -2$, and $x = -3$. We can solve for the associated equilibrium solutions of y by substituting these into equation (2) to obtain the following equilibrium solutions of the system $(x,y) = (1,5)$, $(x,y) = (-2,8)$, and $(x,y) = (-3,9)$.

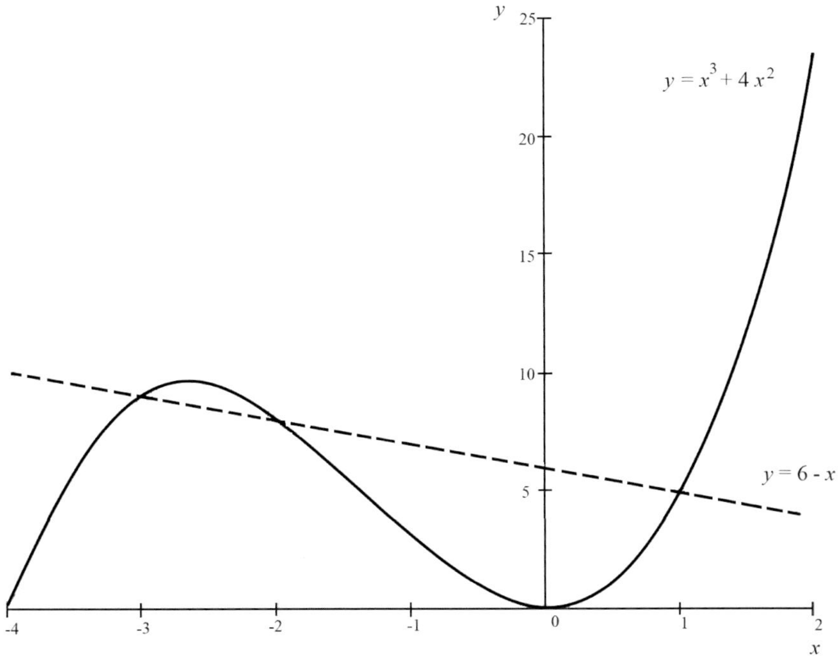

FIGURE 3.5 Solutions for cubic system of equations.

It may be possible to limit the number of solutions to a nonlinear system of equations by either placing restrictions on the system or by noting certain properties of the equations. For example, consider the nonlinear demand and supply system defined by the equations

$$\begin{aligned} p &= q^{-2} \qquad (1) \\ q &= \frac{3}{2} + 2p \qquad (2). \end{aligned} \tag{3.8}$$

Equation (1) is a demand curve with p representing price and q representing quantity. Since quantity can never be negative and $q = 0$ is not defined, the domain for (1) is given by the positive real numbers. It follows that, over the domain of the function, the curve defined by (1) is always downward-sloping, and, since (2) is always upward-sloping in (p, q) space, these equations can only intersect once. Both curves are illustrated in Figure 3.6, which demonstrates the single point of intersection.

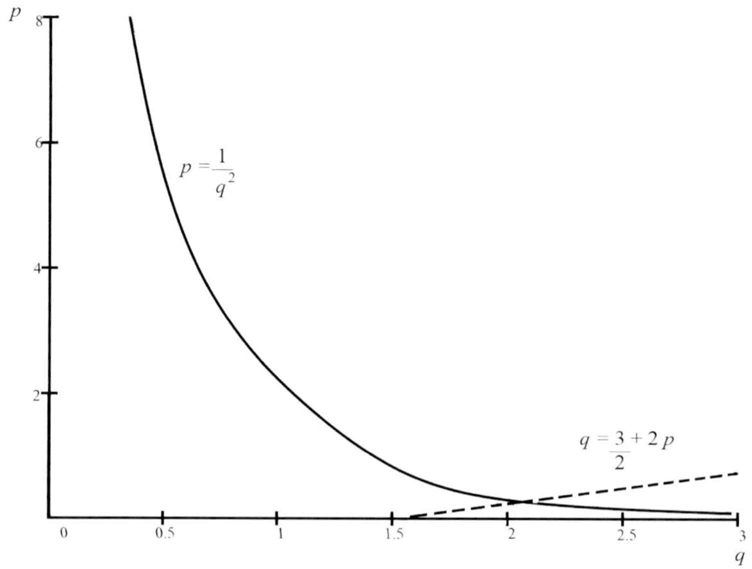

FIGURE 3.6 Nonlinear demand-supply system with unique equilibrium.

In this case, the system can be solved easily using the method of substitution. Substituting the demand equation into the supply curve gives us a single equation of the form $q = 3/2 + 2q^{-2}$. Next, multiplying through by q^2 and rearranging gives us a cubic equation of the form $q^3 - (3/2)q^2 - 2 = 0$. This equation has three roots but, as we have shown, only one of these will have positive values for q and p. In this case, it is easy to see by inspection that $q = 2$ satisfies our equation which, in turn, allows us to solve for p as $p = 1/4$.

REVIEW EXERCISES – SECTION 3.4

1. Find the solutions for the following pairs of nonlinear simultaneous equations.

(a) $y = x^2 - 4x + 6$
$y = x$

(b) $y = x^2 - 4x + 8$
$y = 4x - 8$

(c) $y = x^3 - x^2 + x - 2$
$y = 3x^2 - 4x$

2. The inverse demand curve for a product is given by the equation $p = 4/q$ where $q > 0$. If the supply curve is given by the equation $q = 2 + 2p$, find the values of p and q at which the market is in equilibrium and show that this equilibrium is unique.

3.5 NUMERICAL METHODS

In this section, we show how computer algorithms can be used to solve simultaneous equation models using iterative methods. We discuss two, the Jacobi and Gauss-Seidel methods, and show how these can be applied to very general systems.

The numerical methods which we discuss in this chapter can be applied to both linear and nonlinear systems of equations. However, it is easier to describe them using the example of a linear system. We will begin by setting out the problem of interest as the system of two linear simultaneous equations in two unknown variables x and y shown in

$$\begin{aligned} a_{11}x + a_{12}y = b_1 \quad (1) \\ a_{21}x + a_{22}y = b_2 \quad (2) \end{aligned} . \tag{3.9}$$

The unknown variables of this system are the x and y variables. The parameters are the a and b coefficients. Note that each a coefficient has two subscripts; the first subscript tells us to which equation the parameter belongs, while the second indicates to which variable it is attached. Thus, the parameter a_{12} is the coefficient attached to the second variable (y) in the first equation. The b coefficients are the intercepts for the equations and only require a single subscript which simply tells us to which equation the intercept belongs. We assume that the parameters of the system are known and that we wish to solve for the values of the unknown variables x and y for given values of the a and b coefficients. A useful first step is to write the equations in explicit form as

$$\begin{aligned} x = \frac{b_1}{a_{11}} - \frac{a_{12}}{a_{11}}y \qquad (1) \\ y = \frac{b_2}{a_{22}} - \frac{a_{21}}{a_{22}}x \qquad (2) \end{aligned} . \tag{3.10}$$

Now, suppose we make initial guesses for the solution of the system, which we will label as x_0 and y_0 respectively. Using these guesses we can solve the system as separate equations since each equation now contains only one unknown variable. This gives us

$$\begin{aligned} x_1 &= \frac{b_1}{a_{11}} - \frac{a_{12}}{a_{11}} y_0 \qquad (1) \\ y_1 &= \frac{b_2}{a_{22}} - \frac{a_{21}}{a_{22}} x_0 \qquad (2) \end{aligned} \tag{3.11}$$

This is much easier to solve than a simultaneous equation system, but, unless our initial guesses happened to be the correct solution, it would not give us the answer we want. However, our solution (x_1, y_1) will, under certain conditions, be closer to the true solution than our original guess (x_0, y_0).

If our solution is closer than our original guess, then this suggests a method for solving the system. We can replace the initial guess values with our solution and solve the system again to obtain a new solution (x_2, y_2). The new solution should be even closer to the true solution. We can repeat this process again and again, until the answers we get from solving the equations individually converge on the true solution. This procedure is known as the *Jacobi method,* and the recurrence formulas for the model variables take the following general form

$$\begin{aligned} x_k &= \frac{b_1}{a_{11}} - \frac{a_{12}}{a_{11}} y_{k-1} \qquad (1) \\ y_k &= \frac{b_2}{a_{22}} - \frac{a_{21}}{a_{22}} x_{k-1} \qquad (2) \\ k &= 1, 2, \ldots, K. \end{aligned} \tag{3.12}$$

The accuracy of the solution increases as we increase the number of iterations K. In practice, the number of iterations K is determined by a *convergence criterion*. That is, we stop repeating the calculations when the change in x_k and y_k relative to the previous iteration is sufficiently small.

So far, we have assumed that each iteration gets us closer to the solution, that is that the process will converge. Now, there is no guarantee that convergence will be achieved for all systems of equation. A sufficient, but not necessary, condition for convergence is that the system is *diagonally dominant*. This condition can be stated formally as $|a_{ii}| > \sum_{j \neq i} |a_{ij}|$ for all values of i. Convergence is guaranteed if this condition holds; however, it is possible that the system may converge even if this condition fails.

Another algorithm for solving systems of simultaneous equations is the *Gauss–Seidel method*. This modifies the Jacobi method by making use of intermediate calculations. For example, in, we can replace x_{k-1} in the second equation with x_k. The use of intermediate calculations will generally result in faster convergence than is the case for the Jacobi method. Diagonal dominance is again a sufficient but not necessary condition for convergence when this method is applied. In cases where diagonal dominance is not satisfied, a re-ordering of the equations in the system can sometimes result in convergence. This can occur because the iterative process is sensitive to the ordering of the system when the Gauss–Seidel method is applied, which is not the case for the Jacobi method.

EXAMPLE

Consider the demand–supply system defined by the equations

$$p + 0.5q = 10$$
$$-0.75p + q = -2.$$

We can solve this system numerically using both the Jacobi and the Gauss–Seidel methods. Our starting guess is $p_0 = 0$ and $q_0 = 0$. Some *Python* code is given in Figure 3.7, which shows the routine for the Gauss–Seidel method. The code for the Jacobi method is identical, except that the equation $y1$ = −2 + 0.75*x1 is replaced with $y1$ = −2 + 0.75*x0.[1] The results are shown in Table 3.1. This system is diagonally dominant and, therefore, in both cases, the system converges on the equilibrium $p = 8$, $q = 4$. However, convergence is faster for the Gauss–Seidel method, which converges to an accuracy of 10^{-5} in 14 iterations. In contrast, the Jacobi method converges in 26 iterations.

---

[1] Note that we use the general notation y and x for the variables in our code. Hence, we solve the system by defining $q = x$ and $p = y$.

```
x0 = 0
y0 = 0

converge = 10**-5
cond = 0
counter = 0

print("Iteration","\t","xvalue","\t","yvalue")
print()

while cond == 0:
    print(counter,"\t","\t","%0.4f" % x0,"\t","%0.4f" % y0)
    x1 = 10-0.5*y0
    y1 = -2+0.75*x1
    d1 = abs(x1-x0)
    d2 = abs(y1-y0)
    if d2<converge and d1<converge:
        cond=1
    x0 = x1
    y0 = y1
    counter = counter+1

print()
print("Solution converged after ",counter-1," iterations")
```

FIGURE 3.7 Python code for solution of linear simultaneous equations by Gauss-Seidel method.

TABLE 3.1 Solution of demand-supply system by Jacobi and Gauss-Seidel methods.

Iteration	Jacobi Method		Gauss-Seidel Method	
	Price	Quantity	Price	Quantity
0	0.0000	0.0000	0.0000	0.0000
1	10.0000	-2.0000	10.0000	5.5000
2	11.0000	5.5000	7.2500	3.4375
3	7.2500	6.2500	8.2813	4.2109
4	6.8750	3.4375	7.8945	3.9209
5	8.2813	3.1562	8.0396	4.0297
10	8.0593	4.0297	7.9997	3.9998
15	7.9979	4.0063	8.0000	4.0000
20	7.9996	3.9998	8.0000	4.0000

So far, we have only applied our numerical algorithms to linear systems of equations. One of the useful features of numerical methods like these, however, is that they can be applied to both linear and nonlinear systems. For example, let us consider the Keynesian income-expenditure model defined in. In this model, the endogenous variables are GDP (Y), consumption expenditure (C), tax receipts (T), and imports (M). The exogenous variables are investment (I), government spending (G), and exports (X). Note that, apart from the first equation, all the equations of this system are nonlinear

$$\begin{aligned} Y &= C + I + G + X - M \\ C &= 0.9(Y - T)^{0.95} \\ T &= 0.2Y^{1.05} \\ M &= 0.25Y^{1.1}. \end{aligned} \tag{3.13}$$

This system of equations would be quite hard to solve using either the method of substitution or the method of elimination because of its nonlinear nature. However, such systems can often be solved easily using the iterative numerical methods we now have available to us. The Python code given in Figure 3.8 allows us to solve this particular set of equations. It sets values for the exogenous variables, initial values for the endogenous variables, and a convergence criterion and then uses an iterative loop to solve for the values of the endogenous variables. The equations set out in the code make use of the Gauss–Seidel method but can be easily modified to the Jacobi method for the purposes of comparison.[2] The results for the Gauss–Seidel method are given in Table 3.2 in which convergence to an accuracy of 10^{-2} is achieved in t12 iterations. The Jacobi method also results in convergence, but in this case, it takes 23 iterations.

---

[2] To solve by the Jacobi method, we would replace the lines of code which define the model with the following:

```
Y1 = C0 + I + G + X-M0
C1 = 0.9*(Y0-T0)**0.95
T1 = 0.2*Y0**1.05
M1 = 0.25*Y0**1.1
```

```
Y0 = 200; C0 = 180; T0 = 30; M0 = 100
G = 50; I = 50; X = 100

converge = 10**-2
cond = 0
counter = 0

while cond == 0:
    print(counter,"\t","%4.2f" % Y0, "\t","%4.2f" % C0,"\t",\
          "%4.2f" % T0,"\t","%4.2f" % M0)
    Y1 = C0+I+G+X-M0
    C1 = 0.9*(Y1-T0)**0.95
    T1 = 0.2*Y1**1.05
    M1 = 0.25*Y1**1.1

    d1 = abs(Y1-Y0); d2= abs(C1-C0); d3=abs(T1-T0); d4 = abs(M1-M0)
    if d1<converge and d2 <converge and d3<converge and d4<converge:
        cond = 1

    Y0=Y1; C0=C1; T0=T1; M0=M1
    counter = counter+1

    if counter>100:
        print("Failed to converge in 100 iterations")
        exit()
```

FIGURE 3.8 Python code for solution of Keynesian Income-Expenditure Model by Gauss-Seidel method.

TABLE 3.2 Gauss-Seidel solution of Keynesian income-expenditure model.

Iteration	GDP	Consumption Expenditures	Tax Receipts	Imports
0	200.00	180.00	30.00	100.00
1	280.00	170.72	74.22	122.97
2	247.75	120.68	65.27	107.49
3	213.19	103.70	55.75	91.12
4	212.58	109.62	55.58	90.83
5	218.80	113.86	57.29	93.75
6	220.11	113.59	57.65	94.37
7	219.22	112.77	57.41	93.95
8	218.82	112.66	57.29	93.76
9	218.90	112.79	57.32	93.80
10	218.99	112.84	57.34	93.84
11	218.99	112.82	57.34	93.84
12	218.98	112.81	57.34	93.84

REVIEW EXERCISES – SECTION 3.5

1. Consider the following system of linear equations

$$x + 0.5y = 4$$
$$y - 0.75x = 2$$

Starting with an initial guess $x_0 = 2$ and $y_0 = 3$, calculate the first five iterations of (a) the Jacobi solution method, and (b) the Gauss–Seidel method. Given that the exact solution is $x = 24/11$ and $y = 40/11$. Calculate the errors in each case and show that the Gauss–Seidel solution is closer to the exact solution.

2. Solve the system $p = 4/q, q = 2 + 2p$, where p and q are positive real numbers, using both the Jacobi and Gauss–Seidel methods, and starting values of $p_0 = 0.5$, $q_0 = 2$ to an accuracy of two decimal places. (You can do this easily using a spreadsheet.)

CHAPTER 4

Derivatives and Differentiation

The analysis of change is central to both Economics and Business. For example, we might be interested in how consumers adjust their spending plans as the relative price of commodities varies, or we might want to model how the level of output in the economy adjusts if the central bank alters the interest rate. The branch of mathematics which deals with the analysis of change is *calculus*. There are two main subfields of calculus which are known as *differential calculus* and *integral calculus*, respectively. You will need to become familiar with both in order to conduct economic and business analysis. In this chapter, we will begin by covering the basics of differential calculus.

4.1 DIFFERENTIAL CALCULUS

> *Differentiation* is the process of finding the rate of change of one variable produced by changes in another variable. Differentiation provides an important mathematical tool in both economic and business theory. Although the theory of differentiation can initially appear quite daunting, the practical rules for its application are quite simple.

Differential calculus is concerned with the process of finding the rate at which one quantity changes in response to changes in another related variable. Consider a function of the form $y = f(x)$, where the domain is some subset of the real numbers. Ideally, we would like to measure the *instantaneous* rate of change of y as x changes. As a first attempt, we can find an approximation for this as $\Delta y / \Delta x$, where $\Delta y = f(x_2) - f(x_1)$ and $\Delta x = x_2 - x_1$. This is the slope of the straight line drawn between two points on the function. Differential calculus starts with an approximation of this form and then looks to determine what happens when the change in x is very small.

Consider the example shown in Figure 4.1. The graph shows the quadratic function $y = f(x) = x^2$, where the domain is the set of real numbers $-\infty < x < \infty$. What does Figure 4.1 tell us about the gradient of this function? First, it is obvious that, unlike the case of the linear function, the gradient is not constant. Second, we can see that gradient varies systematically with the value of the x variable. When x is positive, the gradient is also positive, and, as the value of x increases, the gradient increases. If x is negative, then the gradient is negative and becomes larger (in absolute value) as x becomes more negative. This means that the relationship between the gradient and the value of x is itself a function of x.

Now, suppose we wish to find the instantaneous rate of change at $x = 1$. We can interpret this as the slope of the *tangent line* at this point. The tangent line is the straight line which touches the curve at a particular point rather than cutting it at two different points. As a first approximation, we can consider a finite change in the x variable, say from $x = 1$ to $x = 2$. It is very easy to calculate the slope of the straight line between these two points on the function as $\Delta y / \Delta x = (4-1)/(2-1) = 3$, as shown on the diagram. This is an *interval estimate* of the slope and, as such, does not give us the true value of the tangent at the point $x = 1$. As you can see from the diagram, the interval estimate gives an overestimate of the slope of the tangent line. However, we can get a better approximation by considering a smaller increase in x, say from $x = 1$ to $x = 1.5$. This allows us to calculate a new interval estimate of the slope as $\Delta y / \Delta x = \left(1.5^2 - 1\right)/(1.5 - 1) = 2.5$. This will be closer to the tangent slope but remains an over-estimate. Ideally, we would like to make the change in x infinitely close to zero. Setting $\Delta x = 0$ is, of course, not permissible because dividing by zero is not a valid algebraic operation.

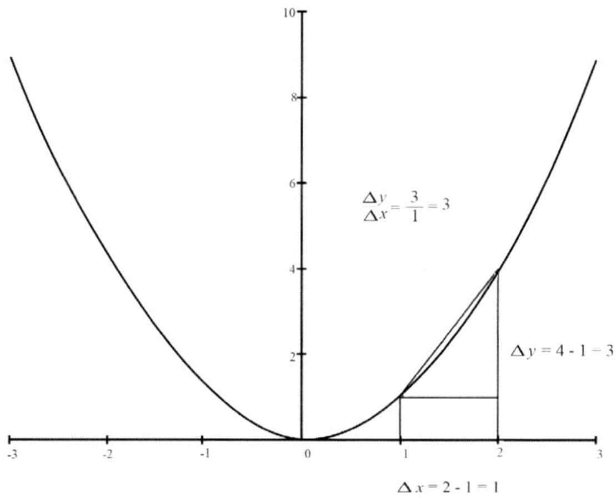

FIGURE 4.1 Interval estimate of the gradient.

Even if we do find the gradient at a particular point, then we are still left with the problem that the slope of a tangent line to a nonlinear function changes as the value of the x variable changes. Rather than looking for a single value of the slope at a point, we need to look for a function of the x variable, which will allow us to determine the value of the gradient at different points in the domain of the function. That is, we need to find a function $f'(x)$ which we will call the *derivative function*. This function may not be defined for all values of x in the domain of the original function. However, the domain of the derivative function $f'(x)$ will always be a subset of the domain of the original function $f(x)$. The process of finding the derivative function is known as *differentiation*.

REVIEW EXERCISES – SECTION 4.1

1. Let $y = f(x) = x^3$ where x is a real number. Calculate interval estimates for the gradient of this function at the following points using a positive increment $\Delta x = 0.01$.

(a) $x = 1$

(b) $x = 2$

(c) $x = 3$

2. Let $y = f(x) = x^3$ where x is a real number. Calculate interval estimates for the gradient of this function at $x = 1$ for the following values of the increment.

(a) $\Delta x = 0.01$

(b) $\Delta x = 0.001$

(c) $\Delta x = 0.0001$

4.2 DIFFERENTIATION FROM FIRST PRINCIPLES

Suppose we have a function $y = f(x)$ which is defined for some subset A of the real numbers. The derivative function is defined as the function $f'(x)$ which gives the slope of the tangent line for different values of $x \in B$ where B is a subset of A. How can we find such a function? The approach we will use here is to construct the derivative function using the infinitesimal numbers which we discussed in Chapter 1. This approach is known as *nonstandard* analysis to distinguish it from the alternative method using limits. The limits approach is referred to as the *standard* approach because it was used to provide the first truly rigorous approach to calculus. However, we have chosen the nonstandard approach here because we believe it is more intuitive and allows us to easily develop many of the important results of differential calculus.

We can define the interval estimate of the gradient of the function $f(x)$ for some interval Δx as

$$\frac{f(x+\Delta x)-f(x)}{\Delta x}. \tag{4.1}$$

Now, suppose Δx is infinitesimal. For a well-behaved (differentiable) function, it follows that the expression (4.1) will be a finite hyperreal number which consists of the sum of a real number (the standard part) and an infinitesimal part. The derivative function is now defined as the *standard part* of (4.1), that is, the remainder, when the infinitesimal part is set equal to zero. We can therefore write our definition of the derivative function as

$$f'(x)=\mathrm{st}\left(\frac{f(x+\Delta x)-f(x)}{\Delta x}\right). \tag{4.2}$$

The derivative function is often indicated using the "prime" notation $f'(x)$. However, an alternative notation, which you will frequently encounter, takes the form dy/dx. Note that we use lower case d to distinguish the derivative, the real valued function, from the interval estimate $\Delta y/\Delta x$, the hyperreal function. The process of finding the derivative function through its definition (4.2) is referred to as *differentiation from first principles*.

EXAMPLE

Consider the function $y=x^2$ and let Δx be a nonzero infinitesimal number. The gradient of the function for an interval equal to Δx is given by

$$\frac{\Delta y}{\Delta x}=\frac{(x+\Delta x)^2-x^2}{\Delta x}=2x+\Delta x. \tag{4.3}$$

Since Δx is infinitesimal, the derivative is given by the expression $f'(x)=dy/dx=\mathrm{st}(\Delta y/\Delta x)=2x$. In this case, the domain of the derivative function is the same as the domain of the original function. That is, both $f(x)$ and $f'(x)$ are defined for all real numbers. This allows us to calculate the gradient of the tangent at any point on the function, that is for any value of x which lies in the open interval $(-\infty,\infty)$. For example, if $x=1$, then the gradient at this point is given by $f'(1)=2$. Similarly, if $x=-2$, then the gradient at this point is $f'(-2)=-4$.

EXAMPLE

Consider the function $y=1/x$. For infinitesimal Δx, we have

$$\frac{\Delta y}{\Delta x}=\frac{1/(x+\Delta x)-1/x}{\Delta x}. \tag{4.4}$$

A little algebra means that we can write this as

$$\frac{\Delta y}{\Delta x}=\frac{1}{\Delta x}\left(\frac{x-(x+\Delta x)}{x(x+\Delta x)}\right)=-\frac{1}{\left(x^2+x\Delta x\right)}.$$

If $x\neq 0$, then the standard part of this expression defines the derivative as $f'(x)=-1/\operatorname{st}\left(x^2+x\Delta x\right)=-1/x^2$. Note that, neither the original function $f(x)$ nor the derivative function $f'(x)$ are defined for $x=0$. The domains of both the original and derivative functions here consist of the set of real numbers which are not equal to zero.

If we can find the derivative of a function for some value of $x=a$, then we say that the function is *differentiable* at this point. For a function to be differentiable at $x=a$, it must be both continuous and smooth at this point. We can think of these conditions intuitively as requiring that the function does not make sudden discrete jumps (continuity) and neither does its rate of change (smoothness). Basically, if we can draw a function without taking the pencil off the page or making sharp changes in the direction in which the pencil travels, then it is likely that it will satisfy these conditions.

A function is *not* differentiable at a point $x=a$ if any of the following are true.

1. $f(a)$ is not defined.
2. $f(a+\Delta x)$ is not defined for some infinitesimal Δx.

3. $\dfrac{f(a+\Delta x)-f(a)}{\Delta x}$ is infinite for some $\Delta x \neq 0$.

4. $\dfrac{f(a+\Delta x)-f(a)}{\Delta x}$ has different standard parts for different infinitesimals $\Delta x \neq 0$.

Let us consider the two example functions we discussed earlier. Using the above criteria, we can show that the function $y = x^2$ is differentiable for all real numbers x. However, the function $y = 1/x$ is not differentiable for $x = 0$ because it is not defined at this point. In fact, there is a discontinuity at $x = 0$ which means that an infinitesimal change in x results in a sudden large change in the value of the function. Any function which has discontinuities will not be differentiable at such points.

Although the presence of discontinuities is one reason why a function may not be differentiable, it is not the only possibility. There are many functions which are continuous but not differentiable at certain points. Let us consider an example.

EXAMPLE

Consider the absolute value function $y = f(x) = |x|$. This is defined for the full set of real numbers $-\infty < x < \infty$. In particular, we note that the function is defined at $x = 0$ where $f(0) = 0$ and that it is continuous at this point since $\operatorname{st} f(\Delta x) = 0$ for all infinitesimal values Δx. Now, consider the derivative function defined by

$$f'(x) = \operatorname{st}\left(\frac{|x+\Delta x| - |x|}{\Delta x}\right). \tag{4.5}$$

We have $f'(0) = 1$ if $\Delta x > 0$ and $f'(0) = -1$ if $\Delta x < 0$. Therefore $f'(x)$ has different standard parts for different infinitesimals when $x = 0$. This is not consistent with our conditions for a function to be differentiable at this point. It follows that the derivative function is not defined for $x = 0$ even though the function $f(x)$ is both defined and continuous at this point.

Having defined the derivative of a function, we will now go on to introduce an important theorem knows as the *increment theorem*. Let $y = f(x)$, and let us assume that the derivative $f'(x)$ is defined at a point x. If Δx is infinitesimal, then the increment theorem states that the change in y is given by the expression

$$\Delta y = f'(x)\Delta x + \varepsilon\Delta x \tag{4.6}$$

where ε is an infinitesimal quantity that depends on x and Δx. This theorem has many applications in the nonstandard approach to calculus and will allow us to derive some important results.

Proof: The proof here follows from the definition of the derivative. We have $f'(x) = \mathrm{st}(\Delta y / \Delta x)$. Any deviation of $\Delta y / \Delta x$ from $f'(x)$ is infinitesimal. Let us label this deviation as ε so that $\Delta y / \Delta x = f'(x) + \varepsilon$, then multiplying through by Δx gives the desired expression $\Delta y = f'(x)\Delta x + \varepsilon\Delta x$.

EXAMPLE

Consider the function $y = x^2$ where x is any real number. We have $\Delta y = 2x\Delta x + (\Delta x)^2$ and, by the increment theorem, we have $\Delta y = 2x\Delta x + \varepsilon\Delta x$. It follows that $\varepsilon = \Delta x$ in this case.

EXAMPLE

Consider the function $y = 1/x$ where x is any nonzero real number. In this case we have $\Delta y = \left(\frac{1}{x+\Delta x} - \frac{1}{x}\right)\Delta x$ or $\Delta y = -\frac{\Delta x}{x(x+\Delta x)}$. From the increment theorem we have $\Delta y = -\frac{1}{x^2}\Delta x + \varepsilon\Delta x$. Setting these equal, and solving for ε gives us $\varepsilon = \frac{\Delta x}{x^2(x+\Delta x)}$.

Using the increment theorem, we define the *differential* of y as $dy = f'(x)\Delta x$. We interpret this expression as the increment in y resulting from an infinitesimal change in x along the tangent line to the function at point x. Note that the differential of x at this point is just equal to the change in x, i.e., $dx = \Delta x$, and, therefore, we can write the differential of y as $dy = f'(x)dx$. The concept of the differential also exists in standard calculus, but it is easier to interpret using the nonstandard approach where dy and dx are infinitesimal changes. The relationship between the differential and the increment in y is illustrated in Figure 4.2.

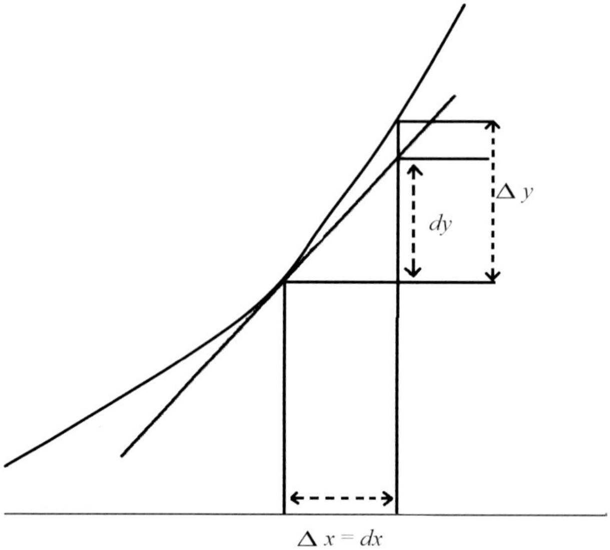

FIGURE 4.2 Relationship between increment in *y* and the differential.

To complete this section, we will consider one final example of finding a derivative from first principles. This is the particularly important example of the exponential function which we introduced in Chapter 2. The exponential function (with the natural base) has the unique property that it is its own derivative, that is, if $y = e^x$, then we also have $dy/dx = e^x$. This somewhat surprising property can be proved as follows:

Proposition: If $y = e^x$ for $-\infty < x < \infty$, then $dy/dx = e^x$ for $-\infty < x < \infty$.

Proof: By the definition of the derivative function, we have

$$\frac{dy}{dx} = \text{st}\left(\frac{e^{x+\Delta x} - e^x}{\Delta x}\right), \tag{4.7}$$

where Δx is infinitesimal. Using this definition, we can write

$$\frac{dy}{dx} = e^x \,\text{st}\left(\frac{e^{\Delta x} - 1}{\Delta x}\right).$$

Recall that the exponential function can be represented as a power series of the form

$$e^{\Delta x} = 1 + \Delta x + \frac{(\Delta x)^2}{2!} + \frac{(\Delta x)^3}{3!} + \ldots$$

It follows that

$$\frac{e^{\Delta x} - 1}{\Delta x} = 1 + \frac{\Delta x}{2!} + \frac{(\Delta x)^2}{3!} + \ldots$$

Since Δx is infinitesimal, it follows that $\mathrm{st}\left(\frac{e^{\Delta x} - 1}{\Delta x}\right) = 1$, and therefore $dy / dx = e^x$. This is a unique property of the exponential function and is one of the reasons why it is so prominent in many areas of mathematics.

REVIEW EXERCISES – SECTION 4.2

1. From first principles, show that the derivative of the function $y = \sqrt{x}$, where x is a positive real number, is given by $1 / \left(2\sqrt{x}\right)$.
2. From first principles, show that the derivative of the function $y = 1 / x^2$, where x is a nonzero real number, is equal to $-2 / x^3$.

4.3 RULES FOR DIFFERENTIATION

It would be very time-consuming to differentiate every function of interest by first principles. Therefore, we develop a set of *rules for differentiation*, which can be applied across a wide range of functions of interest.

The *rules of differentiation* provide a set of results that allow us to find the derivatives of many functions without needing to use first principles. Since the method of first principles is not always easy to apply, these rules can save us a great deal of time and effort. Therefore, we will therefore set out some of the more important rules below, along with proofs and examples where it is useful.

Rule 1: Multiplication by a Constant

Consider a function defined by the equation $u = f(x)$ and let $y = au = af(x)$, where a is a real number. The derivative of y with respect to x is equal to the derivative of u with respect to x multiplied by the same constant, that is $dy/dx = a\,du/dx$.

Proof: This rule is easily proved using the increment theorem. Since $y = af(x)$, we have $\Delta y = af'(x) + \varepsilon\Delta x$, where ε is infinitesimal. The derivative is therefore

$$\frac{dy}{dx} = \text{st}\left(\frac{\Delta y}{\Delta x}\right) = \text{st}\left(af'(x) + \varepsilon\right) = af'(x) = a\frac{du}{dx}.$$

EXAMPLE

We have already shown that, for $u = x^2$, $du/dx = 2x$. Therefore, if we define a new function of the form $y = 2x^2$, it follows that $dy/dx = 4x$.

Rule 2: Sum–Difference Rule

Let $u = u(x)$ and $v = v(x)$. If we now define a new function as the sum, or difference, of these functions, that is either $y = u(x) + v(x)$ or $y = u(x) - v(x)$, then the derivative of this new function will be either the sum or the difference of the derivatives of the original function. That is, if $y = u(x) + v(x)$, then $dy/dx = du/dx + dv/dx$, and if $y = u(x) - v(x)$, then $dy/dx = du/dx - dv/dx$. The proof of this rule is obvious and is left as an exercise for the interested reader.

EXAMPLE

If $y = 4x^2 - 2/x$, then by the sum–difference rule $dy/dx = 8x + 2/x^2$.

Rule 3: The Product Rule

Let $u = u(x)$ and $v = v(x)$. Let $y = f(x) = u(x)v(x)$ then the derivative of y with respect to x is given by the following expression

$$\frac{dy}{dx} = u(x)\frac{dv}{dx} + v(x)\frac{du}{dx}.$$

The proof of this rule is a little trickier than that for the sum-difference rule and is set out explicitly below.

Proof: Let Δx be an infinitesimal change in the x variable. We have

$$\Delta y = (u+\Delta u)(v+\Delta v) - uv = u\Delta v + v\Delta u + \Delta u\Delta v$$
$$\Rightarrow \frac{\Delta y}{\Delta x} = u\frac{\Delta v}{\Delta x} + v\frac{\Delta u}{\Delta x} + \Delta u\frac{\Delta v}{\Delta x}.$$

Since $\Delta v / \Delta x$ and $\Delta u / \Delta x$ have nonzero standard parts but the standard part of $\Delta u \times \Delta v / \Delta x$ is equal to zero, taking the standard part of this expression yields

$$\frac{dy}{dx} = \mathrm{st}\left(\frac{\Delta y}{\Delta x}\right) = u\frac{dv}{dx} + v\frac{du}{dx}$$

which establishes the desired result. This is referred to as the *product rule of differentiation*.

EXAMPLE

Let $y = xe^x$. Defining $u(x) = x$ and $v(x) = e^x$ allows us to use the product rule to find the derivative. We have

$$\frac{dy}{dx} = x\frac{dv}{dx} + e^x\frac{du}{dx} = xe^x + e^x = (x+1)e^x.$$

Rule 4: The Quotient Rule

Let $u = u(x)$ and $v = v(x)$. If we define a new function as $y = f(x) = u(x)/v(x)$, then the derivative of this function is given by the following expression

$$\frac{dy}{dx} = \frac{v(x)\,du/dx - u(x)\,dv/dx}{v(x)^2} \qquad \text{if } v(x) \neq 0.$$

This is not an obvious result but it straightforward to prove as we demonstrate below

Proof: If Δx is an infinitesimal change in the x variable and Δu and Δv are the associated infinitesimal changes in u and v, then we have

$$\Delta y = \frac{u+\Delta u}{v+\Delta v} - \frac{u}{v} = \frac{v(u+\Delta u) - u(v+\Delta v)}{v(v+\Delta v)} = \frac{v\Delta u - u\Delta v}{v^2 + v\Delta v}$$
$$\Rightarrow \frac{\Delta y}{\Delta x} = \frac{v\Delta u / \Delta x - u\Delta v / \Delta x}{v^2 + v\Delta v}.$$

The derivative can now be found by taking the standard part of this expression, which yields

$$\frac{dy}{dx} = \frac{\mathrm{st}(v\Delta u / \Delta x - u\Delta v / \Delta x)}{\mathrm{st}(v^2 + v\Delta v)} = \frac{v\,du/dx - v\,du/dx}{v^2}.$$

which is the desired result.

EXAMPLE

Let $y = e^{-x} = 1/e^x$. Defining $u(x) = 1$ and $v(x) = e^x$ allows us to use the quotient rule to write

$$\frac{dy}{dx} = \frac{e^x du/dx - 1 dv/dx}{\left(e^x\right)^2} = \frac{e^x.0 - 1.e^x}{e^{2x}} = -\frac{1}{e^x}.$$

Rule 5: The Power Function Rule

The Power Function Rule is possibly the most important rule so far. Let $y = x^n$ where x is a real number and n is one of the natural numbers. In this case, the derivative function can be shown to be

$$\frac{dy}{dx} = nx^{n-1}. \tag{4.8}$$

Proof: The proof of this statement uses the method of induction. We first prove that if

$$\frac{dx^{n-1}}{dx} = (n-1)x^{n-2} \tag{4.9}$$

is true, then this implies that (4.8) is true. We then show that this statement is true for $n = 1$, which establishes that it is true for all natural numbers $n = 1, 2, \ldots$.

To establish that the first statement is true, we note that we can write $x^n = x \times x^{n-1}$ and use the product rule to write

$$\frac{dx^n}{dx} = x\frac{dx^{n-1}}{dx} + x^{n-1}\frac{dx}{dx}.$$

If (4.9) is true, then we can write this as

$$\frac{dx^n}{dx} = x(n-2)x^{n-2} + x^{n-1} = (n-1)x^{n-1}.$$

Therefore, it follows that if (4.9) is true, then (4.8) is also true. Now if $n = 1$ then (4.8) is obviously true because $dx / dx = 1 = 1 \times x^0$, and it follows that this statement is true for all natural numbers.

We can extend this result further to include functions of the form $y = x^r$, where r is any real number, but we will need some further results before this is possible. Therefore, we will leave this to the end of this section.

EXAMPLE

For the cubic function $y = x^3$ where x is a real number, we have $dy / dx = 3x^2$.

Note that this establishes a general pattern in that, if the original function is a power function of order n, then the derivative function has order n-1. An important special case here is that of the linear function $y = a + bx$. The derivative of this function is a constant value b which is equal to the slope, or gradient, of the original function.

Rule 6: The Chain Rule

Suppose we have functions $y = f(u)$ and $u = g(x)$, if dy / du and du / dx exist for some value of x, then the *chain rule* states that

$$\frac{dy}{dx} = f'(u)u'(x).$$

Proof: Using the increment theorem, we can write

$$\begin{aligned}\Delta y &= f'(u)\Delta u + \varepsilon_1 \Delta u \\ \Delta u &= g'(x)\Delta x + \varepsilon_2 \Delta x\end{aligned}.$$

where ε_1 and ε_2 are infinitesimal. Combining these expressions yields

$$\Delta y = f'(u)\left[g'(x)\Delta x + \varepsilon_2 \Delta x\right] + \varepsilon_1 \Delta u$$
$$\Rightarrow \frac{\Delta y}{\Delta x} = f'(u)g'(x) + \varepsilon_2 + \varepsilon_1 \frac{\Delta u}{\Delta x}.$$

and taking the standard part of this expression gives the derivative function as

$$\frac{dy}{dx} = \text{st}\left(\frac{\Delta y}{\Delta x}\right) = f'(u)g'(x)$$

which is the required result.

EXAMPLE:

Let $y = \left(2x^2 + 3x\right)^8$. First let us define $u = 2x^2 + 3x$. Given this we have $dy / du = 8u^7$ and $du / dx = 4x + 3$. We can now use the chain rule to find the derivative of the original function by taking the product of these two derivatives which gives us

$$\frac{dy}{dx} = \frac{dy}{du}\frac{du}{dx} = (32x + 24)\left(2x^2 + 3\right)^7.$$

Note that we could have differentiated this function by first expanding the expression and then differentiating the resulting polynomial. However, the polynomial expansion would be very lengthy.

Rule 7: The Inverse Function Rule

Suppose the function $y = f(x)$ has inverse function $x = g(y)$, then the derivative of x with respect to y is given by

$$\frac{dx}{dy} = \frac{1}{dy / dx}.$$

Proof: Using the increment theorem, we have $\Delta y = f'(x)\Delta x + \varepsilon \Delta x$. Dividing both sides by Δy gives us

$$1 = f'(x)\frac{\Delta x}{\Delta y} + \varepsilon \frac{\Delta x}{\Delta y}$$

and rearranging yields

$$\frac{\Delta x}{\Delta y} = \frac{1}{f'(x) + \varepsilon}.$$

Taking the standard part defines the derivative of x with respect to y as

$$\frac{dx}{dy} = \text{st}\left(\frac{\Delta x}{\Delta y}\right) = \frac{1}{f'(x)}.$$

EXAMPLE

Let $y = x^2$ where $x \geq 0$. This has inverse function $x = \sqrt{y}$ where y has domain equal to the nonnegative real numbers. Since $dy / dx = 2x$, it follows from the inverse function rule that

$$\frac{dx}{dy} = \frac{1}{dy / dx} = \frac{1}{2x} = \frac{1}{2\sqrt{y}}.$$

One very important application of the inverse function rule is to determine the derivative of the log function. We can demonstrate this as follows

Proposition: The derivative of the log function $y = \ln(x)$ is equal to $dy / dx = 1 / x$.

Proof: Let $y = \ln(x)$ where x is a positive real number. We wish to find dy / dx. The inverse function is $x = e^y$, which has domain equal to the set of real numbers. Recall that we have already found the derivative of the exponential function as $dx / dy = e^y$. Therefore, by the inverse function rule, we have

$$\frac{dy}{dx} = \frac{1}{dx / dy} = \frac{1}{e^y} = \frac{1}{x}.$$

As an aside, we note that, if $y = \ln x$, then the differential of the log function gives us $dy = (1 / x) dx$. This is an important result which is frequently used when working with logarithmic functions.

Generalization of the Power Function Rule

Rules 1 to 6 will allow us to find the derivatives of many of the functions we encounter. Before we go any further, however, we need to generalize the power function rule to cases in which the exponent is a real number. In our earlier proof, we showed that, if x is a real number, and n is a natural number, the derivative of the function $y = x^n$ is given by the expression $dy / dx = nx^{n-1}$. We will now demonstrate that this remains true when the exponent is any real number.

To show this, let $y = x^r$ where r is a real number. We have $\ln y = r \ln x$ and, by the definition of the differential, we have

$$\frac{1}{y} dy = r \frac{d \ln x}{dx} = \frac{r}{x} dx .$$

Rearranging this expression gives us

$$\frac{dy}{dx} = r \frac{y}{x} = r \frac{x^r}{x} = rx^{r-1} .$$

Note that this holds for all real numbers r, not just the natural numbers. We can therefore use the Power Function Rule to differentiate any function of the form $y = x^r$, where both x and r are real numbers.

REVIEW EXERCISES – SECTION 4.3

1. Find the derivative of the function $y = 4x(x+1)^2$ using the product rule.
2. Find the derivative of the function $y = (3x-1)/(x+2)^2$ using the quotient rule.
3. Find the derivative of the function $y = \sqrt{(4x^2 + 2x)}$ using the chain rule.
4. Find the derivative of the function $y = \left(\sqrt{x}\right)^{4/5}$ using the power function rule.

4.4 SOME ECONOMIC EXAMPLES

> Differentiation provides an important mathematical tool for microeconomic theory. In this section, we show how the derivative function can be used to analyze the properties of the demand curve. The two issues we consider are the calculation of the marginal revenue function and the price elasticity of demand.

Consider a firm facing a downward sloping inverse demand curve which takes the general form $p = a - bq$, where p is price, q is quantity, and a and b are parameters that are assumed to be positive. The total revenue from sales is equal to the product of price and quantity. We can therefore write an equation for total revenue of the form

$$R(q) = aq - bq^2. \tag{4.10}$$

Since the inverse demand curve is linear in quantity, it follows that the total revenue function is quadratic. Marginal revenue is defined as the increase in revenue from a small increase in quantity sold. It follows that the marginal revenue function can be calculated as the derivative of the total revenue function. We have

$$MR = \frac{dR(q)}{dq} = a - 2bq. \tag{4.11}$$

Marginal revenue is therefore also a linear function of quantity. However, the gradient of this function is different from that of the inverse demand function. The slope of the inverse demand function is given by the parameter $-b$, whereas that of the marginal revenue function is equal to $-2b$.

The inverse demand and marginal revenue functions are plotted in Figure 4.3 where the parameters are $a = 1$ and $b = 1/2$. The inverse demand curve therefore has the equation $p = 1 - 0.5q$, and the marginal revenue function has equation $MR = 1 - q$. The two equations have the same intercept on the vertical axis where $p = MR = 1$ when $q = 0$. The inverse demand function cuts the horizontal axis at the point $q = 2$, while the marginal revenue function cuts the horizontal axis halfway between this point and the origin at the point $q = 1$. It follows that, in the interval $0 \le q < 1$, both price and marginal

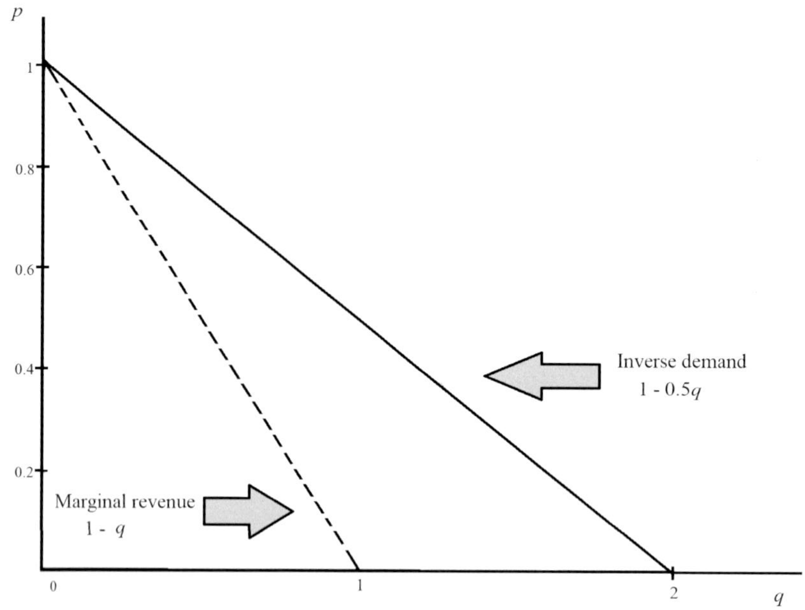

FIGURE 4.3 Inverse demand and marginal revenue functions.

revenue are positive. That is, for values of q in this range, the firm can increase its revenue by increasing output. In the range $1 < q \leq 2$, marginal revenue is negative, even though price remains positive. In this range therefore, the firm can increase revenue by cutting output, with the increase in price more than offsetting the loss of revenue due to a reduction in sales. Intuitively therefore, the point $q = 1$, which corresponds to a value of $p = 0.5$, is the value of output at which the firm's revenue is maximized. This is confirmed by the graph of the total revenue function shown in Figure 4.4, which indicates a maximum point when $q = 1$.

The derivative can also be used to calculate the *price elasticity of demand*. This is a measure of the responsiveness of quantity demanded to a change in price. It is defined as minus one multiplied by the percentage change in quantity demanded divided by the percentage change in price. It can be written as

$$\eta^D = -\frac{\Delta q / q}{\Delta p / p} = -\frac{\Delta q}{\Delta p}\frac{p}{q}. \tag{4.12}$$

The expression given in (4.12) is the *arc elasticity*, that is, the response in demand measured between two distinct two points on the demand curve.

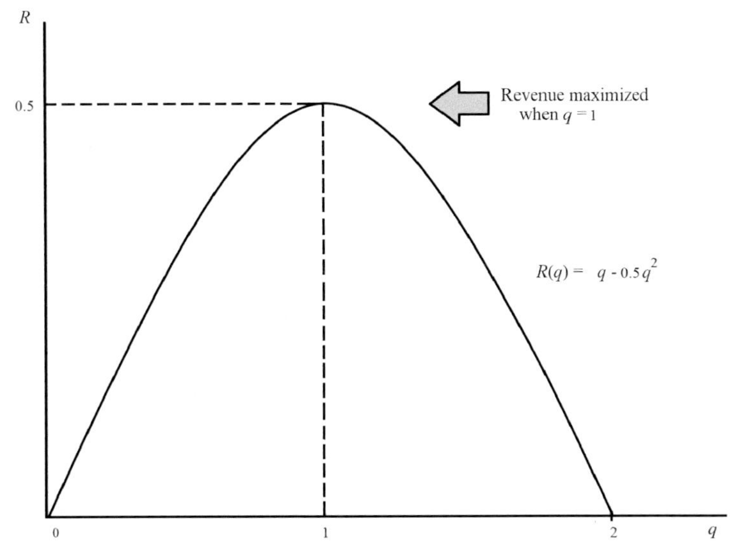

FIGURE 4.4 Total revenue function.

Since we normally expect demand to be negatively related to price, the convention is to multiply the elasticity by minus one to express it as a positive number. The larger the value of the elasticity, then the more responsive quantity demanded is to price. If we wish to measure elasticity at any point, then we can replace the term $\Delta q / \Delta p$ with the derivative dq / dp. If the elasticity is greater than one, then we say that demand is *price elastic*, if the elasticity is less than one, then we say that demand is *price inelastic*.

EXAMPLE

Consider the linear demand function $q = 100 - 2p$. Given that price and quantity must each be greater than or equal to zero, the domain of this function is $0 \leq p \leq 50$, and the range is $0 \leq q \leq 100$. The inverse demand function can be derived as $p = 50 - 0.5q$. The point elasticity of demand is given by the expression

$$\begin{aligned} \eta^D &= -\frac{dq}{dp}\frac{p}{q} = -(-2)\times\frac{50-0.5q}{q} \\ &= \frac{100}{q} - 1 \end{aligned}$$

The value of the elasticity of demand is therefore a declining function of the level of output. q can take on values in the closed interval $[0,100]$. Let us consider the value of q at the end points of this interval. When $q = 100$ and $p = 0$, the elasticity is equal to zero. If $q = 0$, then the elasticity is not defined. However, we can say that, as $q \to 0$, the elasticity tends to infinity. For $q = 50$, we have $\eta^D = 1$, and therefore for $0 \le q < 50$, the elasticity is greater than one and marginal revenue is positive while for $50 < q \le 100$, the elasticity is less than one and marginal revenue is negative.

The examples we have considered demonstrate that the elasticity of demand is not constant along a linear demand curve. However, we can choose an alternative functional form for which the elasticity of demand remains constant at all points on the curve. Consider the function

$$q = ap^{-b} \tag{4.13}$$

where a and b are both positive parameters. Using the power function rule for differentiation, we have $dq/dp = -bap^{-b-1}$, and since p can only take on values greater than or equal to zero, it follows that the gradient of this curve is always less than or equal to zero. Suppose we now think of (4.13) as a demand curve and calculate the elasticity of demand. This is given by

$$\eta^D = -\frac{dq}{dp}\frac{p}{q} = -1 \times \left(-bap^{-b-1}\right) \times \frac{p}{ap^{-b}} = b.$$

That is, the elasticity of demand for this demand curve is constant and given by the parameter b.

EXAMPLE

Consider the function $q = 50p^{-2}$. The price elasticity of demand for this function is equal to the value of the exponent, that is 2. The graph of this function is shown in Figure 4.5. This shows that the function has asymptotes given by the horizontal axis, where $q \to 0$ as $p \to \infty$, and the vertical axis, where $q \to \infty$ as $p \to 0$.

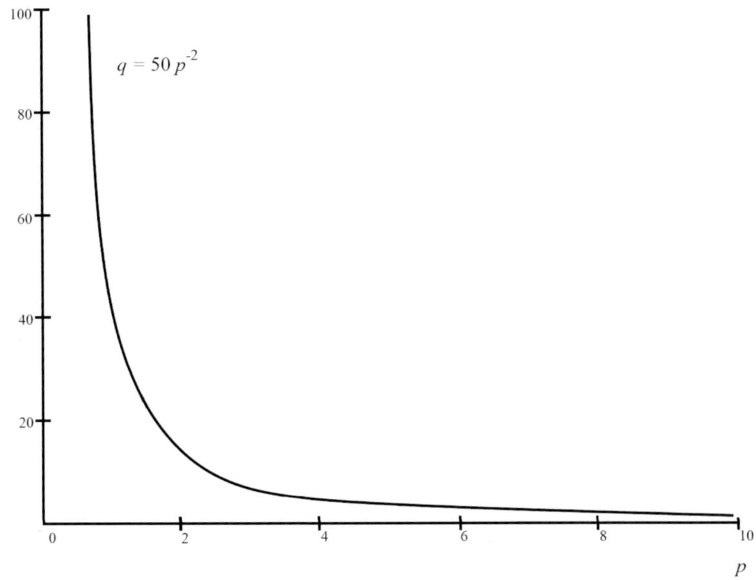

FIGURE 4.5 Graph of constant elasticity demand curve.

REVIEW EXERCISES – SECTION 4.4

1. Consider the linear demand function $q = 60 - 3p$. Find an expression for the price elasticity of demand as a function of output and, using this expression, find the range of values of output for which demand is price elastic and for which it is price inelastic.
2. Consider the demand curve $q = 20p^{-5}$. Find the derivative of the demand function and, using this, show that the slope of the demand curve tends to zero as $p \to \infty$ and to minus infinity as $p \to 0$.

4.5 HIGHER-ORDER DERIVATIVES

The derivative is itself a function of the independent variable x and can be differentiated to find *higher-order derivatives*. Such derivatives contain important information about the curvature of the function under consideration and prove useful when looking for *turning points* in functions, which will be covered in Chapter 5.

When we differentiate a function of the form $y = f(x)$, then we generate a new function of x which takes the form $dy/dx = f'(x)$. This function may itself be differentiable, in which case, we obtain the *second derivative* of y with respect to x, which we write $d^2y/dx^2 = f''(x)$. In general, if it is possible to differentiate the function $y = f(x)$ n times, then the nth order derivative of y with respect to x is written as

$$\frac{d^n y}{dx^n} = f^{(n)}(x).$$

Higher-order derivatives like this are useful when analyzing the properties of functions and when we are looking for the turning points in functions which indicate maximum or minimum points.

EXAMPLE

Consider the polynomial function $y = 4x^3 + 3x^2 + 2x + 1$. This has derivatives

$$\begin{aligned} \frac{dy}{dx} &= f'(x) = 12x^2 + 6x + 2 \\ \frac{d^2y}{dx^2} &= f''(x) = 24x + 6 \\ \frac{d^3y}{dx^3} &= f^{(3)}(x) = 24 \\ \frac{d^n y}{dx^n} &= f^{(n)}(x) = 0 \quad \text{for all } n \geq 4. \end{aligned}$$

The example above illustrates an interesting property of polynomial functions, in that the nth order derivative eventually becomes zero. This is not true for all functions.

EXAMPLE

Consider the function $y = 1/x$, this has derivatives

$$\begin{aligned} \frac{dy}{dx} &= f'(x) = -\frac{1}{x^2} \\ \frac{d^2y}{dx^2} &= f''(x) = \frac{2}{x^3} \end{aligned}$$

$$\frac{d^3y}{dx^3} = f^{(3)}(x) = -\frac{6}{x^4}$$

$$\ldots$$

$$\frac{d^ny}{dx^n} = f^{(n)}(x) = \frac{(-1)^n n!}{x^{n+1}}.$$

This is an example of a *continuously differentiable function* for which the derivatives never become zero. There are many such examples of continuously differentiable functions which are of interest to us.

The definition of higher-order derivatives allows us to introduce the *Taylor series*. This is an important mathematical series that allows us to find approximations to very general functional forms as polynomial series. The Taylor series is defined as follows: any function $y = f(x)$, which is continuously differentiable at a point $x = a$, can be represented as an infinite polynomial function of the form

$$\begin{aligned} f(x) &= f(a) + f'(a)(x-a) + \frac{f''(a)}{2!}(x-a)^2 + \frac{f^{(3)}(a)}{3!}(x-a)^3 + \ldots \\ &= \sum_{n=0}^{\infty} \frac{f^{(n)}(a)}{n!}(x-a)^n . \end{aligned}$$

If we truncate this expression after $m+1$ terms, then we obtain the mth order *Taylor series polynomial*. This can often be used as an approximation to the function which is more easily manipulated than the original function.

EXAMPLE

Consider the function $y = f(x) = 1/x$. The second-order Taylor series polynomial for this function around the point $a = 1$ can be derived as $g(x) = 3 - 3x + x^2$. (This is left as an exercise for the reader.) If, we plot $f(x)$ and $g(x)$ for the range $0 < x < 2$, as shown in Figure 4.6, then we see that the Taylor series polynomial provides a reasonably good approximation to the function when x is close to $a = 1$. However, the approximation becomes progressively worse the further we move away from this point.

Another interesting application of the Taylor series is to the exponential function. An important property of this function is that differentiation simply returns the original function. That is, if $y = \exp(x)$, then dy/dx is also equal

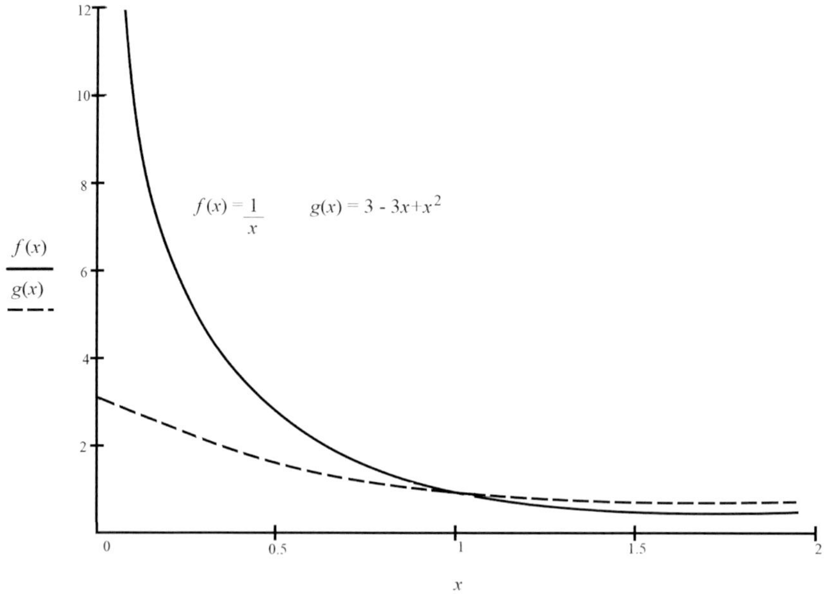

FIGURE 4.6 $y = f(x) = 1/x$ and a second-order Taylor series approximation.

to $\exp(x)$. This means that we can take any order derivative $d^n y / dx^n$, and we will always get the function $\exp(x)$ as the result. Now, let us consider the Taylor series for this function around the point $x = 0$. Since $\exp(0) = 1$, we have

$$\exp(x) = 1 + x + \frac{x^2}{2!} + \frac{x^3}{3!} + \ldots = \sum_{n=0}^{\infty} \frac{x^n}{n!}.$$

The higher-order terms in this expression will tend to zero because $n!$ tends to infinity faster than x^n.[1] Thus, we can approximate the exponential function using a finite-order polynomial function, which can be very useful for some problems.

The Taylor series can also be applied to the log function. For $-1 < x \le 1$, we have

$$\ln(1+x) = x - \frac{x^2}{2} + \frac{x^3}{3} - \frac{x^4}{4} + \ldots = \sum_{n=1}^{\infty} \frac{(-1)^{n-1} x^n}{n}.$$

[1]Note that this provides a proof that the representation of the exponential function which we introduced in Chapter 2 is valid.

The proof of this result is one of the exercises for this section. The approximation $\ln(1+x) \approx x$ for small values of x is often convenient in the analysis of growth over time.

REVIEW EXERCISES – SECTION 4.5

1. Find the first and second derivatives of the following functions

(a) $y = -\frac{a}{x^2}, \ x > 0$

(b) $y = \exp(2x), \ -\infty < x < \infty$

(c) $y = 3\ln(x), \ x > 0$

2. Find the Taylor series expansion of the function $y = \ln(1+x)$ around $x = 0$ and show that this gives a convergent sequence when $-1 < x < 1$.

4.6 NUMERICAL METHODS

Analytical differentiation is not always easy, or even possible. In this section, we develop some numerical methods for calculating the derivative of a function which can be used when analytical methods fail.

Numerical methods for calculating derivatives are based around *finite differencing* methods. That is, we take a small interval h and calculated an estimate of the derivate based on this interval. We can calculate estimates based on a *forward difference* of the form

$$f'(x) \approx \frac{f(x+h) - f(x)}{h} \tag{4.14}$$

or a *backward difference* of the form

$$f'(x) \approx \frac{f(x) - f(x-h)}{h}. \tag{4.15}$$

We can often improve on both these methods however, by using a *central difference* of the form

$$f'(x) \approx \frac{f(x+h/2)-f(x-h/2)}{h}. \tag{4.16}$$

For all these cases, the estimate will be improved by taking the smallest possible interval h to calculate the derivative. At some stage however, we run up against the constraint that computers can only calculate numbers to a limited degree of accuracy. For computers that store numbers in double precision format, this means that we are limited to calculations based on numbers smaller than 10^{-15}. In most practical situations, this means that we can calculate estimates of derivatives to a reasonably high degree of accuracy.

One way to improve on the accuracy of the derivative estimate for a given interval size is to make use of the *Richardson Extrapolation*. The error magnitude for estimates based on the standard method (4.16) is $O(h^2)$. That is the error is proportional to the square of the step-size. Therefore if $h = 0.01$, then the error will be of magnitude 10^{-4}. Now, we can define two alternative central difference estimators as

$$\begin{aligned} D_1(h) &= \left(\frac{1}{2h}\right)\left(f(x+h)-f(x-h)\right) \\ D_2\left(\frac{h}{2}\right) &= \left(\frac{1}{h}\right)\left(f\left(x+\frac{h}{2}\right)-f\left(x-\frac{h}{2}\right)\right). \end{aligned} \tag{4.17}$$

Each of these will have errors of the same order of magnitude $O(h^2)$. However, we can define a linear combination of these estimates which has error magnitude $O(h^4)$. Thus, for example, if $h = 0.01$, then the order of magnitude of the error in the estimate will be 10^{-8}. This linear combination takes the form shown in equation (4.18). The code in Figure 4.7 allows us to assess the relative accuracy of these methods.

$$D(h) = \frac{4D_2(h/2)-D_1(h)}{3}. \tag{4.18}$$

The code in Figure 4.7 is designed to a calculate the derivative of the function $y = \exp(x)$ at $x = 1$ based on an interval length of $h = 0.01$. The analytical derivative for this function is known and is equal to $\exp(1)$ at this point. Therefore, we can assess the accuracy of our estimates on this basis. Using this code, we obtain the output shown in Table 4.1. This illustrates the increase in accuracy from the use of the Richardson extrapolation.

```
import math

def f(x):
    f = math.exp(x)
    return f

def riche(x,h):
    d1 = (f(x+h)-f(x-h))/(2*h)
    d2 = (f(x+h/2)-f(x-h/2))/h
    riche = (4*d2-d1)/3
    return riche

def mpd(x,h):
    mpd = (f(x+h/2)-f(x-h/2))/h
    return mpd

h = 0.01
x = 1
deriv = math.exp(x)

print("Estimates")
print()
print("Central difference method\t",round(mpd(x,h),6))
print("Richardson extrapolation\t",round(riche(x,h),6))
print()
print("Errors")
print()
print("Central difference method\t",abs(mpd(x,h)-deriv))
print("Richardson extrapolation\t",abs(riche(x,h)-deriv))
```

FIGURE 4.7 Code for numerical estimates of derivative for function $y = \exp(x)$.

TABLE 4.1 Alternative numerical estimates of the derivative of the exponential function.

Estimates	
Central difference method	2.718293
Richardson extrapolation	2.718282

Errors	
Central difference method	1.1326×10^{-5}
Richardson extrapolation	5.6691×10^{-11}

REVIEW EXERCISES – SECTION 4.6

1. Show that the truncation error for the forward difference estimate of the derivative as given by equation (4.14) is $O(h)$.
2. Using the code provided, compare the accuracy of the central difference estimate and the Richardson extrapolation estimate for the following derivatives.

   (a) $f(x)=x^3$ at $x=2$.

   (b) $f(x)=\ln(x)$ at $x=1$.

CHAPTER 5

OPTIMIZATION

When we model the behavior of agents such as households and firms, we often come across the problem of locating the maximum or minimum points of functions. For example, we may wish to find the level of output or price level which maximizes the profits of a firm, the consumption of a good or product that maximizes the utility of a household, or the mix of labor and capital inputs that minimizes costs of production. Differential calculus provides a crucial mathematical tool for this purpose.

5.1 IDENTIFYING CRITICAL POINTS

> Critical points identify candidates for a maximum or minimum point of a function. These occur at points where the first derivative of a function is equal to zero, it is not defined, or at the end points of its domain.

Suppose we have a function $f(x)$ where x is a real number which is defined on a closed interval with lower limit a and upper limit b. The *extreme value theorem* states that such a function has both a maximum and a minimum value. $x_{\min}$ corresponds to a minimum if $f(x) \geq f(x_{\min})$ and $x_{\max}$ corresponds to a maximum if $f(x) \leq f(x_{\max})$ for all possible values of the variable x. These conditions define the *global minimum* $f(x_{\min})$ and the *global maximum* $f(x_{\max})$ of the function. Differential calculus provides us with a powerful mathematical tool we can use to find these critical points. Note that the global maximum and minimum points may not be unique, it is possible that the function may attain its maximum and minimum values at several different values of x.

Our first task is to identify a set of candidate points, and then to determine which of these correspond to maximum or minimum values. Values of x which generate candidates for maximum or minimum points are referred to as *critical values*. The *critical point theorem* states that, to be a maximum or minimum point, $x = c$ must satisfy one of the following three conditions:

1. $f'(c) = \left.\frac{dy}{dx}\right|_{x=c} = 0$
2. $f'(c) = \left.\frac{dy}{dx}\right|_{x=c}$ is not defined
3. c is an end point. That is, either $c = a$ or $c = b$.

Let us consider each of our conditions in turn. Take the condition $f'(c) = 0$. Points that satisfy this condition are referred to as *stationary points*. This condition captures a situation in which the function "flattens out" at some point in the interior of its domain. For a *local minimum*, this would appear as when a function that was decreasing flattens out and starts to increase. For a *local maximum*, a function that had been increasing becomes flat and then starts to decrease. Note the qualification *local* in these cases because it is possible that there may be multiple points that have the property $f'(c) = 0$ and the global minimum or maximum may occur at any of these or at points that correspond to conditions (2) or (3). The condition $f'(c) = 0$ may not even indicate a local maximum or minimum. A third possibility, in this case, is that function flattens out and then starts to move again in its previous direction. This is referred to as a *point of inflexion*. The use of the condition $f'(x) = 0$ to locate a possible turning point is referred to as a *first-order condition* because it identifies a candidate point based on the first derivative, but it does not tell us what type of point we have located.

As an example, consider the function $f(x) = x^2 - 3x$ where $0 \le x \le 2$. The first derivative of this function is $f'(x) = 2x - 3$ which is zero when $x = 3/2$, indicating that $x = 3/2$ is a critical value. The value of the function at this point is $f(3/2) = -9/4$. This is a local minimum because values of $f(x)$ in the vicinity of $x = c$ are all greater than this value. This can be demonstrated easily because $f(3/2 + \delta) = -9/4 + \delta^2$ for nonzero values of δ. It follows immediately that this is a local minimum. In fact, this point satisfies the conditions for a global minimum because the derivative function is defined for all values of x in the domain, so no additional critical points arise through the second condition, and the values of the function at the end points are $f(0) = 0$ and $f(2) = -2$ which are both greater than the value at the turning point.

Returning to the general case, we can illustrate the three possible types of stationary point corresponding to the condition $f'(x)=0$ using the graphs shown in Figures 5.1 (a), (b), and (c).

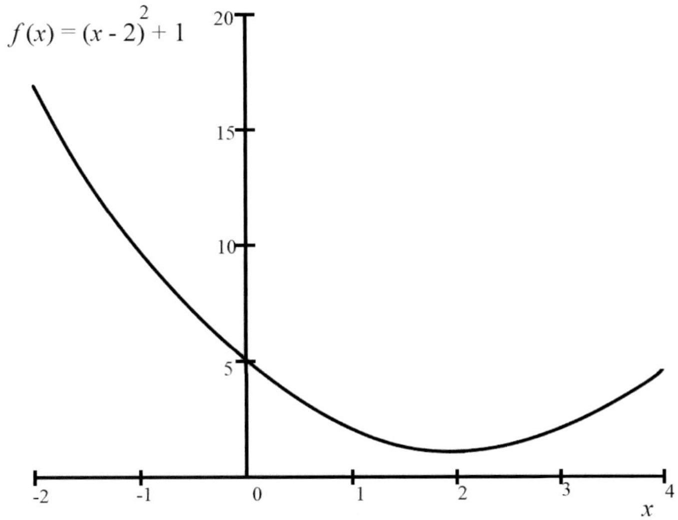

FIGURE 5.1(a) Function with local minimum at $x = 2$.

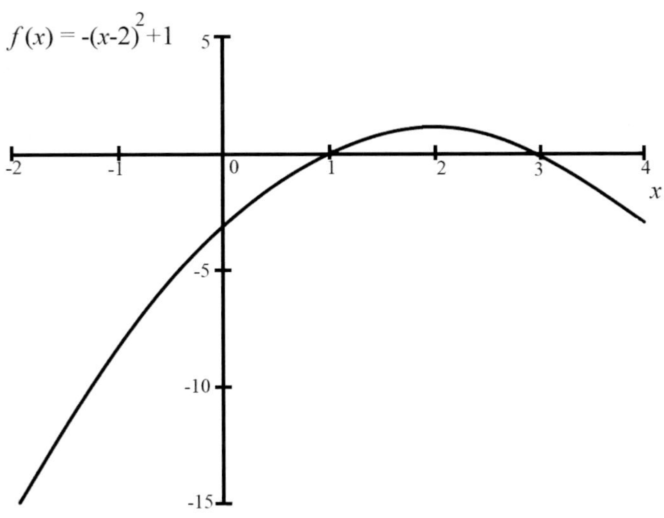

FIGURE 5.1(b) Function with local maximum at $x = 2$.

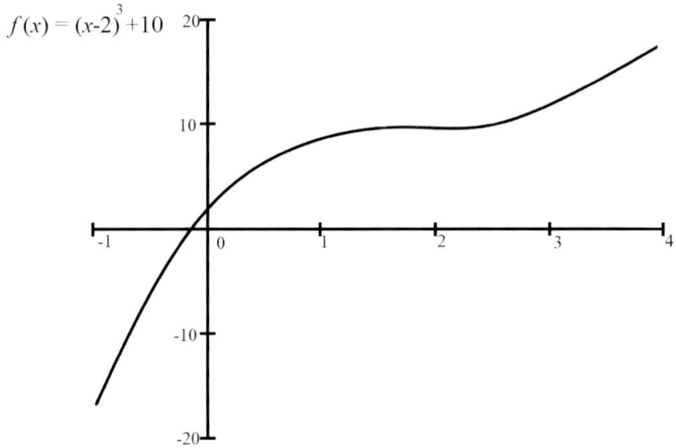

FIGURE 5.1(c) Function with point of inflexion at $x = 2$.

Identifying the nature of stationary points is made easier by using *second-order conditions*. The first-order condition is simply the requirement that $f'(x) = 0$ when $x = c$. The second-order condition relies on the second-order derivative of the function at this point. For a local minimum point like the one we have identified in the example, the derivative will switch sign from negative to positive at $x = c$. Hence, a sufficient condition for a point to be a local maximum is $f''(c) > 0$, or that the function is *concave up* at this point. Similarly, at a local maximum, the sign of the derivative will switch from positive to negative, and $f''(c) < 0$, referred to as *concave down*, is therefore a sufficient condition for a point to be a local maximum. If, however, $f''(c) = 0$, then the second-order condition fails to identify the nature of the turning point. Such a point may be either a local maximum, minimum, or a point of inflexion.

EXAMPLE

Find, and identify, all the critical points of the function $y = f(x) = x^3/3 + x^2/2 - 2x$, where x lies in the interval $-3 \le x \le 2$.

The first derivative of this function is

$$f'(x) = \frac{dy}{dx} = x^2 + x - 2.$$

This is continuous and differentiable on the interval $-3 \le x \le 2$. Therefore, there are no critical points at which $f'(x)$ is not defined. For interior solutions, we look for values of x such that $f'(x) = 0$. Factorizing the expression for the first derivative and setting this equal to zero gives

$$f'(x) = (x-1)(x+2) = 0.$$

Therefore, the two possible solutions are $x = 1$ and $x = -2$, which both lie within the domain. The second-order derivative is

$$f''(x) = \frac{d^2y}{dx^2} = 2x + 1.$$

At $x = 1$, we have $f''(1) = 3 > 0$, and therefore this is a local minimum. The value of the function at this point is $f(1) = -7/6$. At $x = -2$, we have $f''(-2) = -3 < 0$, and therefore this is a local maximum, and the value of the function at this point is $f(-2) = 4/3$.

Next, we check the end points of the function. We have $f(-3) = 3/2$ and $f(2) = 2/3$. Both of these points are greater than the local minimum we have identified and less than the local maximum. It follows that the local minimum we have identified at $x = 1$ is also the global minimum, and the local maximum we have identified at $x = -2$ is also the global maximum. These properties are confirmed by inspection of the graph of the function which is given in Figure 5.2.

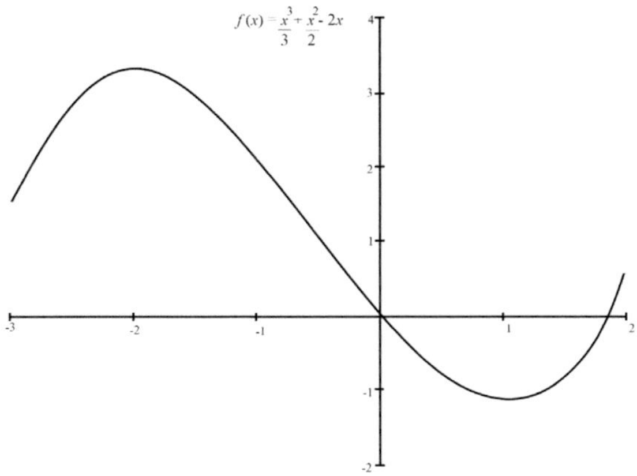

FIGURE 5.2 Graph of the function $f(x) = x^3/3 + x^2/2 + 2x$.

If the second derivative is equal to zero at a stationary point, then we must rely on other conditions to establish its nature. Such points can be either a local maximum, a local minimum, or a point of inflexion.

EXAMPLE

Consider the function $f(x) = x^3$ where x is a real number. This function has first and second-order derivatives $f'(x) = 3x^2$ and $f''(x) = 6x$. It follows that there is a stationary point at $x = 0$ because $f'(0) = 0$, but this cannot be identified using the second-order condition as $f''(0) = 0$. However, for small changes in x around the stationary point equal to δ, we have $f'(\delta) = 3\delta^2$. Thus, $f'(\delta) > 0$ for both positive *and* negative values of δ. Since the derivative does not change sign around the stationary point, it follows that this is a point of inflexion rather than a local maximum or local minimum.

EXAMPLE

Consider the function $f(x) = x^4$ where x is a real number. This function has first and second derivatives $f'(x) = 4x^3$ and $f''(x) = 12x^2$. We have $f'(0) = 0$ and therefore, a stationary point at $x = 0$, but we also have $f''(0) = 0$, and again the second-order condition does not tell us the nature of this point. However, it is easy to establish that this is a local minimum by direct inspection of the first derivative function. For a small change in x equal to δ, we have $f'(\delta) = 4\delta^3$. This is positive when $\delta > 0$ and negative when $\delta < 0$. It follows that the derivative changes sign from negative to positive around this point, which is enough to demonstrate that this is a local minimum.

So far, we have assumed that the function under consideration is defined on a closed interval. This simplifies things because it allows us to evaluate the function at the end points and compare these directly with any interior stationary points when we look for global minimum or maximum points. This becomes a bit trickier when the function we consider is defined on an open interval. In these cases, we need to modify our definition of the global minimum and maximum points and introduce the concepts of the *infimum* and the *supremum*.

EXAMPLE

Consider the function $f(x) = 3x^2 + x + 2$ defined on the closed interval $-1 \leq x \leq 1$. The first-order and second-order conditions identify a local minimum at the point $x = -1/6$, and this is also the global minimum with $f(-1/6) = 23/12$. Evaluating the function at the end points gives $f(1) = 6$ and $f(-1) = 4$. Therefore, the global maximum for the function occurs at the end point $x = 1$ with $f(1) = 6$.

Now, consider the same equation, but with the domain of x redefined as the open interval $-1 < x < 1$. $x = 1$ is no longer part of the domain of this function, and therefore there is no value of x such that $f(x) = 6$, so this point can no longer be defined as the global maximum of the function. It remains the case, however, that we can choose values of x which are arbitrarily close to 1 and which therefore generate values of $f(x)$ which are arbitrarily close to 6. In this case, we say that the supremum of the function is equal to 6.

The *supremum* of a function is therefore defined as the smallest real number s such that $f(x) < s$ for all values of x in the domain. This is a generalization of the idea of the global maximum, which allows for cases in which the function is defined on an open interval. A related concept is that of the *infimum*, which is the greatest real number l such that $f(x) > l$, for all values of x in the domain. Again, this can be thought of as a generalization of the idea of the global minimum to cases in which the function is defined on an open interval.

EXAMPLE

Consider the function $f(x) = 3x^3 - x$ defined on the open interval $-1 < x < 1$. A plot of this function is given in Figure 5.3.

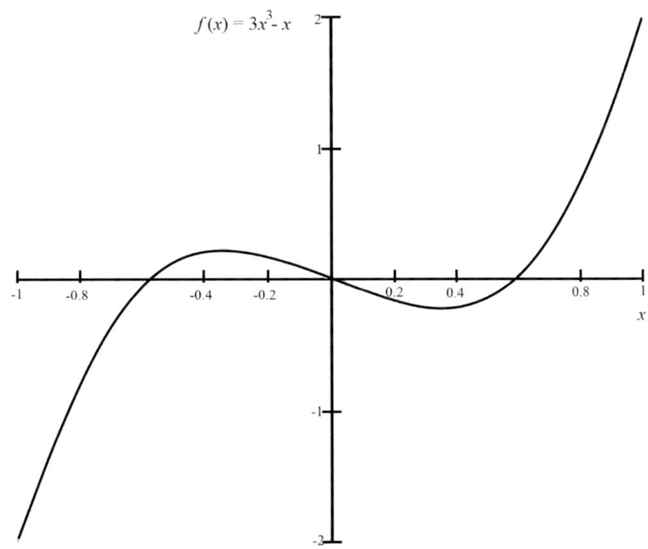

FIGURE 5.3 Plot of the function $f(x) = 3x^3 - x$; $-1 < x < 1$.

This function has a local maximum at the point $x=-1/3$, and a local minimum at the point $x=1/3$. Neither of these points, however, correspond to either a supremum of an infimum of the function since there are clearly values of x that give a higher value for the function than $f(-1/3)=2/9$, and values of x which give a lower value than $f(1/3)=-2/9$.

As x approaches 1 from below, the value of the function approaches 2, but we cannot say that this is the global maximum value of the function because $x=1$ is not part of the domain. Instead, we say that 2 is the supremum of the function because it is the lowest real number such that $f(x)<2$ for all values of x in the domain. Similarly, as x approaches the value -1 from above, the value of the function approaches -2, but this cannot be called the global minimum of the function because $x=-1$ is not part of the domain. In this case, we say that -2 is the infimum of the function because it is the largest real number such that $f(x)>-2$ for all values of x in the domain.

REVIEW EXERCISES – SECTION 5.1

1. Find, and identify, all critical points for the following functions

(a) $f(x)=4x^2+2x \qquad -1\le x\le 2$

(b) $f(x)=x^3-12x \qquad -5\le x\le 5$

(c) $f(x)=\frac{2}{3}x^3-2x \qquad -2\le x\le 2$

2. Find the interior critical points for the following functions and determine whether they are maximum or minimum points

(a) $f(x)=x\ln(x) \qquad 0<x<\infty$

(b) $f(x)=2/(x^2+1) \qquad -\infty<x<\infty$

(c) $f(x)=3|x|-x \qquad -\infty<x<\infty$

3. Show that the function $f(x)=1/x;\ 1\le x<\infty$ has global maximum value equal to one and infimum zero.

5.2 SOME ECONOMIC EXAMPLES

> Microeconomics is the study of the behavior of agents looking for the best possible solutions given limited resources. For example, we may be concerned with agents who decide on consumption patterns to maximize utility or firms who decide on output levels to maximize profits. Calculus provides the tools to formalize such decision making and, in this section, we show how it can be applied to a variety of economic problems.

In this section, we look at how we can use the first and second-order derivative conditions for turning points in the context of microeconomic theory. Our first example concerns the profit-maximizing decision of a firm.

Consider a firm that faces a downward-sloping demand curve $p = a - bq$, and has costs which are determined by the function $C = cq$, where q is the level of output. a, b, and c are all positive. Here, the parameters are the intercept and slope coefficients of the demand curve and the slope coefficient of the cost function. To find the profit-maximizing level of output for this firm, we set up the profit function as $\Pi(q) = R(q) - C(q)$ where R and C are revenue and costs of production, both of which are functions of the level of output q. Using the demand curve and the cost function, we have

$$\begin{aligned} \Pi(q) &= (a - bq)q - cq \\ &= (a - c)q - bq^2. \end{aligned} \tag{5.1}$$

We note that the domain of this function is given by the range of values of q, which are consistent with price and quantity, both being nonnegative. Thus, we have $0 \le q \le a / b$.

The first-order condition for a maximum is found by differentiating with respect to q and setting this derivative equal to zero. This gives

$$\Pi'(q) = (a - c) - 2bq = 0. \tag{5.2}$$

Therefore, there is a stationary point at $q = (a - c) / 2b$. For this to be positive, we need $a > c$. If $a \le c$, then there is no level of output which will yield positive profits. To confirm that the critical point we have identified is a local maximum, we check the second-order condition. The second-order derivative is equal to $-2b$, which is negative because of the assumption that b is positive.

Finally, we check for other possible critical points. The first derivative is always defined on the domain, and therefore there are no critical points corresponding to $f'(x)$ being undefined. At the end points of the function, we have $\Pi(0)=0$ and $\Pi(a/b)=-ac/b$. If $a>c$ then there is a level of output $q=(a-c)/2b>0$ which generates positive $\Pi(q)=(a-c)^2/4b$ and this is greater than the value of the function at either of the end points. Therefore, under the assumption that $a>c$, there is a unique local maximum of the function corresponding to the condition $\Pi'(q)=0$, and this is also the global maximum for the function.

EXAMPLE

Let the parameters of the model take the following values: $a=1$, $b=0.5$ and $c=0.5$ the profit function now takes the form $\Pi(q)=0.5q-0.5q^2$, and its first derivative is given by the expression $\Pi'(q)=0.5-q$. For a maximum, we require $\Pi'(q)=0$, which gives $q=0.5$. We can confirm that this is a maximum by checking the second-order condition, which is satisfied in this case because $\Pi''(q)=-1<0$. The profit function for this problem is shown in Figure 5.4, which confirms the existence of a maximum point at $q=0.5$.

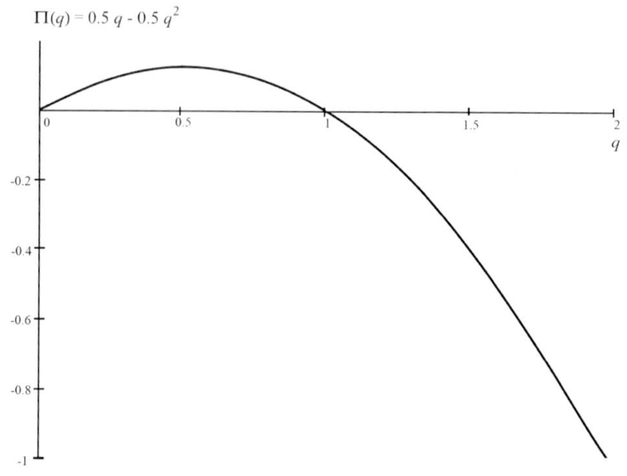

FIGURE 5.4 Graph of the profit function for $a=1$, $b=0.5$, and $c=0.5$.

Let us consider another example based on the analysis of costs of production. In general, we can make a distinction between fixed costs F, which are independent of the level of production, and variable costs, $V(q)$ which depend on how much output the firm chooses. We can therefore write the total cost function as

$$TC(q) = F + V(q). \tag{5.3}$$

The average cost function is equal to the total cost divided by the level of output. Therefore, we have

$$AC(q) = \frac{F}{q} + \frac{V(q)}{q}. \tag{5.4}$$

Note that these functions are very general. We can be a little more specific by assuming that variable costs increase as the level of output increases. This means that the derivative of the variable cost function is positive, that is, $V'(q) > 0$. Under this assumption, we can demonstrate the very general result that the average cost of production is minimized when the marginal cost $V'(q)$ is equal to the average cost.

To demonstrate this result, we use the first-order condition for a minimum. Differentiating with respect to output using the quotient rule and setting the derivative equal to zero gives us the condition shown in equation (5.5).

$$-\frac{F}{q^2} + \frac{qV'(q) - V(q)}{q^2} = 0. \tag{5.5}$$

Rearranging this expression gives

$$-\frac{1}{q}\left(\frac{F}{q} - V'(q) + \frac{V(q)}{q}\right) = 0. \tag{5.6}$$

For a local minimum, we need the term in parentheses to equal zero, which gives the condition

$$V'(q) = \frac{F + V(q)}{q}. \tag{5.7}$$

The left-hand side of this equation is the derivative of the variable cost function with respect to output, that is, the *marginal cost*. The right-hand side is equal to the sum of fixed plus variable costs divided by the level of output, that is, the average cost of production. We have therefore demonstrated our desired result, that is, for a local minimum, the marginal cost of production must equal the average costs of production. This is a very general result, which does not depend on the form taken by the cost function.

EXAMPLE

Consider the total cost function $TC = 100 + 5q + 4q^2$ where $q \geq 0$. The marginal cost is found by differentiating this function to give $MC = 5 + 8q$. The average cost function is obtained by dividing total cost by output to obtain $AC = 100/q + 5 + 4q$. To find the level of output at which average cost is minimized, we differentiate the average cost function and solve for the value of output at which the derivative is equal to zero.

$$\frac{dAC}{dq} = -\frac{100}{q^2} + 4 = 0 \Rightarrow q = 5.$$

Note that there are two roots for this equation $q = \pm\, 5$. We discard the negative root because it does not lie within the domain of the function. When $q = 5$, we have $MC = 5 + 8 \times 5 = 45$, and $AC = 100/5 + 5 + 4 \times 5 = 45$. Therefore, marginal and average costs are equal at the cost-minimizing level of output. The relationship between the average and marginal cost functions is shown in Figure 5.5.

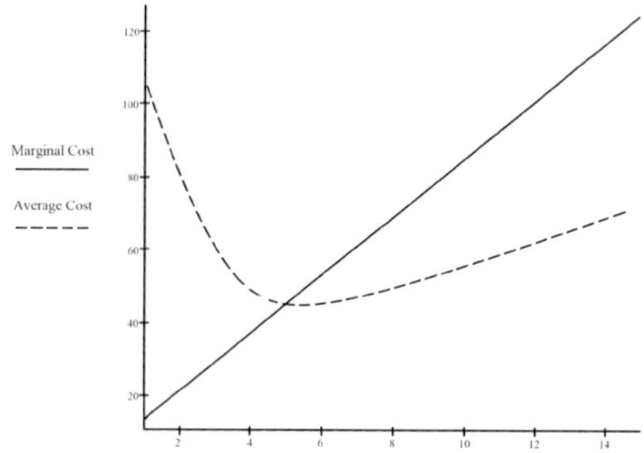

FIGURE 5.5 Marginal and average costs.

Next, let us consider a slightly more complicated example drawn from the theory of consumption. Suppose we have an individual with a fixed endowment of money seeking to maximize utility by spreading consumption expenditure across two time periods. We will assume a utility function of the form

$$u = c_1^a + \frac{1}{1+\delta} c_2^a, \tag{5.8}$$

where c_1 and c_2 are consumption in periods 1 and 2 respectively, a is a parameter which we assume lies in the range 0 to 1, and δ is the rate of time discount. If $\delta > 0$, then this function assumes that the consumer puts a higher weight on current consumption relative to consumption in the future.

The budget constraint is given by

$$c_1 + \frac{1}{1+r}c_2 = M, \tag{5.9}$$

where M is the initial endowment of money available to the consumer and r is the interest rate. This budget constraint assumes that the agent can borrow or lend freely at the market interest rate r. Using equation (5.9) we can transform the problem from one in which there are two choice variables, c_1 and c_2, to one in which there is a single choice variable, c_1. The first-order condition for a maximum therefore becomes

$$\frac{dc_1^a}{dc_1} + \frac{d}{dc_2}\left(\frac{c_2^a}{1+\delta}\right)\frac{dc_2^a}{dc_1^a} = 0. \tag{5.10}$$

and, since $c_2^a = (1+r)(M - c_1^a)$, this can be written as

$$ac_1^{a-1} - a\frac{1+r}{1+\delta}c_2^{a-1} = 0, \tag{5.11}$$

which can be solved to give the following equation for c_1,

$$c_1 = \left(\frac{1+\delta}{1+r}\right)^{\frac{1}{1-a}} c_2. \tag{5.12}$$

Equation (5.12) has some interesting properties. In particular,

1. If the interest rate is equal to the rate of time discount, $\delta = r$, then $(1+\delta)/(1+r) = 1$, and therefore consumption is equal in both time periods.
2. If the interest rate is less than the rate of time discount, $\delta > r$, then $(1+\delta)/(1+r) > 1$, and the agent consumes more in period 1 than in period 2.
3. If the interest rate is greater than the rate of time discount, $\delta < r$, then $(1+\delta)/(1+r) < 1$, and the agent consumes less in period 1 than in period 2.

This illustrates a very general result in the analysis of intertemporal choice in that we see that the optimum distribution of consumption over time depends on the relationship between the rate of interest and the rate at which agents discount future utility derived from consumption.

EXAMPLE

Let the parameter $a = 0.5$, the rate of time discount equal 0.05, and the market interest rate equal 0.1.

From equation (5.12) we have

$$c_1 = \left(\frac{1.05}{1.1}\right)^2 c_2 = 0.9112\ c_2.$$

Thus consumption is significantly lower in period 1 because the market rate of interest is higher than the rate of time discount.

REVIEW EXERCISES – SECTION 5.2

1. A firm faces the inverse demand curve $p = 72 - 2q$, and its costs of production are given by $C = 10q^2$, where q is output. Find the profit-maximizing level of output using the first derivative condition and show that this is a maximum using the second-order condition.
2. A firm faces inverse demand curve $p = 10/\sqrt{q}$, and its costs of production are given by $C = 5q$, where q is output. Find the profit-maximizing level of output using the first derivative condition and show that this is a maximum using the second-order condition.
3. A firm has a total cost function $TC = 100 + 3q + 4q^2$. Find the level of output which minimizes average cost and show that marginal cost is equal to average cost at this level.

5.3 CONVEXITY AND CONCAVITY

The properties of convexity and concavity refer to the shape of a function. If a function has these properties, then it limits the number of turning points and allows us to determine their nature more easily.

A function is said to be *weakly convex* if the secant line, a line segment drawn between any two points on the function lies on, or above, the function itself. This can be stated formally as follows.

> $f(x)$ is a weakly convex function if $\lambda f(x_1)+(1-\lambda)f(x_2)\geq f(\lambda x_1 + (1-\lambda)x_2)$ where $0\leq\lambda\leq 1$ and x_1 and x_2 are points in the domain.

Similarly, a function is said to be *weakly concave* if the secant line, a line segment drawn between any two points on the function lies on, or below, the function itself, that is

> $f(x)$ is a weakly concave function if $\lambda f(x_1)+(1-\lambda)f(x_2)\leq f(\lambda x_1+(1-\lambda)x_2)$ where $0\leq\lambda\leq 1$ and x_1 and x_2 are points in the domain.

If the inequalities used for the definitions of convexity and concavity hold strictly (except at the end points), then the function is said to be either strictly convex or strictly concave. That is, a strictly convex function has the property $\lambda f(x_1)+(1-\lambda)f(x_2)> f(\lambda x_1+(1-\lambda)x_2)$, and a strictly concave function has the property $\lambda f(x_1)+(1-\lambda)f(x_2)< f(\lambda x_1+(1-\lambda)x_2)$, for $0<\lambda<1$.

We can get an intuitive understanding of these definitions from the examples shown in Figure 5.6. A line drawn between any two points on a strictly convex function will always lie above the function itself, except at the end points. Similarly, a line drawn between any two points on a strictly concave function will always lie below the function itself, except, of course, for the end points. Neither strict convexity nor strict concavity is consistent with a straight-line function. However, a straight line can be said to be simultaneously both weakly convex and weakly concave.

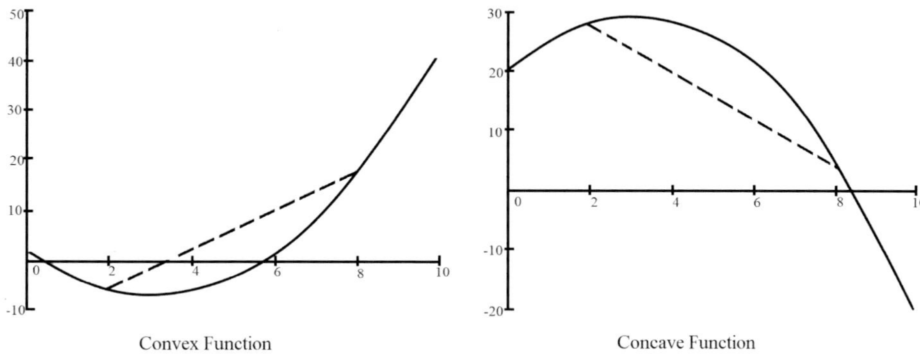

FIGURE 5.6 Strictly convex and strictly concave functions.

The properties of convexity and concavity are of interest to us because they limit the number of maximum or minimum points associated with a function. For example, a convex function can have at most one local minimum, while a concave function can have at most one local maximum.

If a function is twice differentiable, then conditions for strict convexity and concavity can be defined in terms of its second derivative. These can be stated as follows:

1. If the second derivative is positive for all points in the domain, then the function is strictly convex.
2. If the second derivative is negative for all points in the domain, then the function is strictly concave.

The reverse is not true. The fact that a function is strictly convex does not mean that its second derivative is always negative. This can easily be demonstrated with a counter example. The function $y = x^4$, where x is a real number, is strictly convex, as is immediately obvious when the function is plotted. However, the second derivative is equal to $12x^2$ which is equal to zero at $x = 0$.

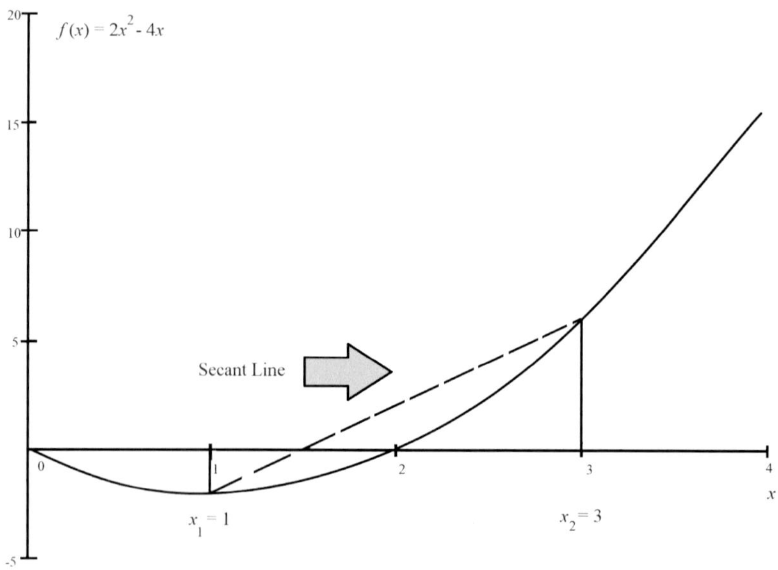

FIGURE 5.7 Convex function example.

We can give a more formal proof using the increment theorem. Let us consider the case of the convex function shown in Figure 5.7. For a strictly convex function, the slope of the secant from point x_1 to x_2 will be greater than the slope of the tangent line at x_1 when $x_2 > x_1$, and less than the slope of the tangent line when $x_2 < x_1$. From the increment theorem, we have

$$\Delta y = f'(x)\Delta x + \varepsilon \Delta x$$

where ε is an infinitesimal number which is a function of x and Δx. Consider a Taylor series expansion of the function $f(x+\Delta x)$ around x_1. We have

$$f(x_1+\Delta x) = f(x_1) + f'(x_1)\Delta x + \frac{f''(x_1)}{2!}(\Delta x)^2 + \frac{f^{(3)}(x_1)}{3!}(\Delta x)^3 + \ldots.$$

Subtracting $y = f(x_1)$ from both sides and dividing by Δx gives us

$$\frac{\Delta y}{\Delta x} = f'(x_1) + \left\{\frac{f''(x_1)}{2!}\Delta x + \frac{f^{(3)}(x_1)}{3!}(\Delta x)^2 + ..\right\}.$$

The term $\Delta y / \Delta x$ is the slope of the secant line and $f'(x_1)$ is the slope of the tangent line at x_1. The term in curly parentheses in the above expression gives us the difference between these quantities. In fact, the term in curly parentheses is the ε term from the increment theorem. Since Δx is infinitesimal, higher powers of Δx can be neglected. If $f''(x_1) > 0$ and $\Delta x > 0$, then ε is positive infinitesimal, and if $f''(x_1) > 0$ and $\Delta x < 0$, then it is negative infinitesimal. Therefore, the difference between the slope of the secant line and the tangent function has the same sign as Δx. It therefore follows that if $f''(x_1) > 0$, the function is strictly convex. By the same argument, if $f''(x_1) < 0$, it follows that the function is strictly concave.

REVIEW EXERCISES – SECTION 5.3

1. Show, from first principles, that the function $y = x^2$ is strictly convex.

2. Using the second-order derivative condition, determine which of the following functions are concave and which are convex.

(a) $f(x) = 1/x \qquad x > 0$

(b) $f(x) = -4x^3 - x^2 \qquad x > 0$

(c) $f(x) = \ln x \qquad x > 0$

5.4 NUMERICAL METHODS FOR FINDING TURNING POINTS

In some cases, it may be difficult to locate turning points using analytical methods. However, numerical methods can often be used to solve such problems relatively easily. These methods usually involve iterative calculations.

In this section, we consider numerical methods for finding turning points of functions. For simplicity, we consider functions that are continuous and differentiable. It is possible to consider numerical methods which relax both assumptions, but the necessary algorithms are considerably more complicated. The advantage of the assumptions is that, if $f(x)$ is differentiable, then we can limit our search to values of x such that $f'(x)=0$, or, in numerical terms, we can look for the roots of the first derivative function. This is a standard problem in numerical analysis, and there are several ways in which we can approach it.

First, we will consider the *bracketing method* for finding roots. Suppose we have two values of x such that $f'(x_L)$ and $f'(x_U)$ have opposite signs. If $f'(x)$ is continuous, the intermediate value theorem tells us that there is some value of x in the interval x_L to x_U at which the derivative is zero, as illustrated in Figure 5.8. To narrow the interval, let us consider a point halfway between these values, that is, $x_M=(x_L+x_U)/2$. As shown in the diagram, this gives a value $f'(x_M)$ with the same sign as $f'(x_U)$. The upper limit of the interval containing the root can therefore be redefined as $x_U=x_M$. If we had found that $f'(x_M)=f'(x_L)$, then we would have redefined the lower limit as $x_L=x_M$. We can continue to repeat this process to obtain progressively narrower intervals until the lower and upper limits have converged to some acceptable limit. For example, we might set an acceptable tolerance limit such that we stop the process when $x_U-x_L<10^{-7}$.

It is not necessary to have an analytical expression for the first derivative function to apply the bracketing method (or other numerical methods, for that matter). Instead, we can use a finite difference method in which we approximate the derivative using an expression of the form

$$f'(x)\approx\frac{f(x+h)-f(x)}{h},$$

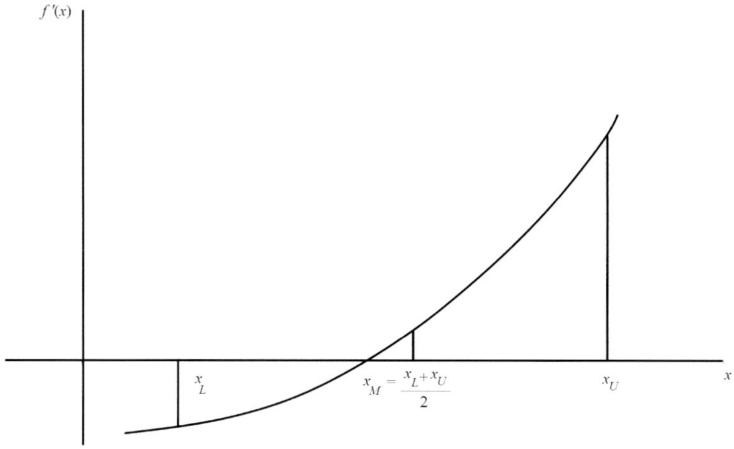

FIGURE 5.8 The bracketing method.

where h is a small increment in the x value. This is the *forward difference* estimate. Alternative estimates are given by the *backward difference* estimate, which takes the form

$$f'(x) \approx \frac{f(x) - f(x-h)}{h},$$

or the *symmetric difference quotient,* which takes the form

$$f'(x) \approx \frac{f(x+h) - f(x-h)}{2h}.$$

This estimate is the slope of the secant line[1] between two points, one just below and one just above the point of interest. The choice of the increment h is also important in determining the accuracy of the estimate. Ideally, we want h to be as close to zero as possible so that the estimate of the secant slope is as close as possible to the slope of the tangent at a point. However, there is a limit to the accuracy of computer calculations as h becomes small. The convention here is to set h to be approximately equal to the cube root of machine epsilon. This is the smallest number ε such that the computer recognizes a difference between 1 and $1+\varepsilon$. For modern computers using double precision arithmetic, machine epsilon is approximately 10^{-16}. This suggests a

[1] A secant line is simply any line which passes through two points on a curve.

```
import math

def f(x):
    f=x*math.exp(-1*x/3)
    return f

def fd(x):
    fd=(f(x+h)-f(x-h))/(2*h)
    return fd

h=10**-5
xlower = 0.0
xupper = 5.0
converge = 10.0**7
counter = 0

if fd(xlower)*fd(xupper)>0:
    print("Derivative has same sign at lower and upper limits")
    exit()

print("Iteration","\t","Lower value","\t","Upper value","\t","Difference")
print()

while abs(converge)>10**-5:
    counter = counter+1
    if counter>100:
        print("Failed to converge after 100 iterations")
        exit()
    xmid = (xlower+xupper)/2
    if fd(xmid)*fd(xlower)>=0:
        xlower = xmid
    if fd(xmid)*fd(xupper)>=0:
        xupper = xmid
    converge=xlower-xupper
    print(counter,"\t","\t","%0.4f" % xlower,"\t","%0.4f" % xupper,"\t","%0.6f" % converge)

print()
print("Solution converged at iteration ",counter)
```

Figure 5.9 Python code for the bracketing algorithm.

value of h of approximately 10^{-5}. This appears to work reasonably well and is, therefore, the value that we will use in all our future calculations.

The code shown in Figure 5.9 implements the algorithm described in the previous paragraphs. The function itself, and the derivative function, are given in the function definitions at the top of the code. The initial upper and lower limits, the value of h, and the convergence criterion are set at the top, with the iterative loop for the search being contained in the *while* loop. The example chosen here is the function $f(x)=x\exp(-x/3)$, which has a maximum at the value $x=3$. Running this code gives the results shown in

Figure 5.10, which demonstrates that the algorithm converges to the correct solution in 19 iterations.

Iteration	Lower Limit	Upper Limit	Difference
1	2.5000	5.0000	-2.5
2	2.5000	3.7500	-1.25
3	2.5000	3.1250	-0.625
4	2.8125	3.1250	-0.3125
5	2.9687	3.1250	-0.15625
6	2.9687	3.0469	-0.07813
7	2.9687	3.0078	-0.03906
8	2.9883	3.0078	-1.95E-02
9	2.9980	3.0078	-9.77E-03
10	2.9980	3.0029	-4.88E-03
11	2.9980	3.0005	-2.44E-03
12	2.9993	3.0005	-1.22E-03
13	2.9999	3.0005	-6.10E-04
14	2.9999	3.0002	-3.05E-04
15	2.9999	3.0000	-1.53E-04
16	3.0000	3.0000	-7.63E-05
17	3.0000	3.0000	-3.81E-05
18	3.0000	3.0000	-1.91E-05
19	3.0000	3.0000	-9.54E-06

FIGURE 5.10 Results of bracketing method search for stationary point.

An alternative, and potentially more efficient, way of locating stationary points is provided by *Newton's method*. Given some initial guess, Newton's method uses a linear approximation to the derivative function to generate an improved estimate for the root of the derivative function. An illustration is given in Figure 5.11. At $x = x_k$, the slope of the derivative function is given by $f''(x_k)$, the second derivative of the original function evaluated at this point. The value of the derivative function is equal to $f'(x_k)$ at this point, and, therefore, we can calculate the point at which the tangent to the derivative function crosses the horizontal axis as

$$x_{k+1} = x_k - \frac{f'(x_k)}{f''(x_k)}. \tag{5.13}$$

This provides a value of x which is closer to the root of the function than the initial guess and repeating the process will generate further estimates which are even closer. Thus, (5.13) provides a recurrence relationship which we can use to iterate toward a solution. Using this relationship, we continue the process until the change in the value of x is less than some predetermined tolerance level. Note that, as with the first derivative, we do not need an analytical expression for the second derivative to implement this method. Instead, we can use an approximation of the form

$$f''(x) \simeq \frac{f(x+h) - 2f(x) + f(x-h)}{h^2}. \qquad (5.14)$$

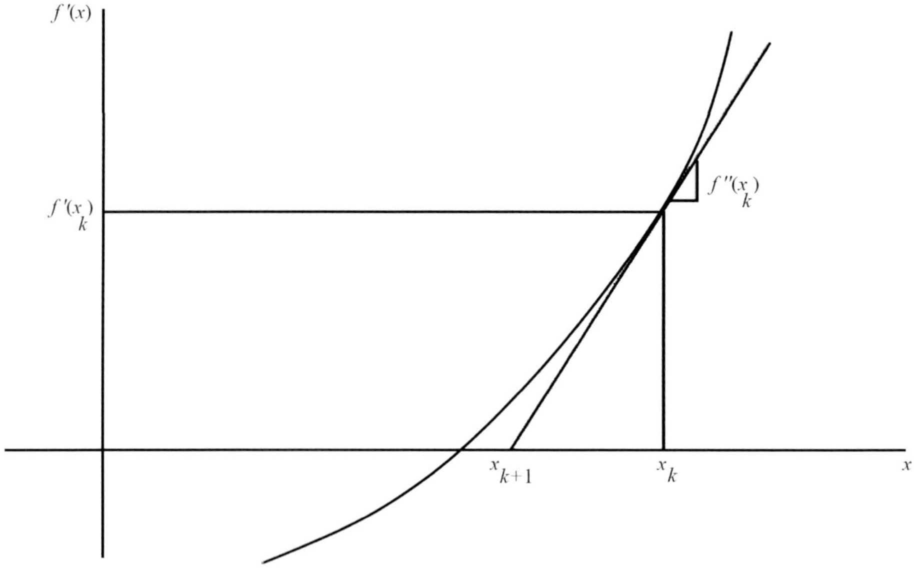

FIGURE 5.11 Newton's method.

The code shown in Figure 5.12 implements Newton's method for the function $f(x) = x\exp(-x/3)$. Although both the first and second derivatives can be calculated explicitly here, this code uses numerical derivatives for the purposes of illustration. Figure 5.13 reports the output from this code. Newton's method shows improved efficiency as it takes only seven iterations to achieve the same level of accuracy as the bracketing method output shown in Figure 5.10, which took 19 iterations to obtain a result within the tolerance level of 10^{-7}. Note that the negative second derivative at the solution immediately identifies this turning point as a maximum rather than a minimum.

```
import math

def f(x):
    f=x*math.exp(-1*x/3)
    return f

def fd(x):
    fd=(f(x+h)-f(x-h))/(2*h)
    return fd

def sd(x):
    sd=(f(x+h)-2*f(x)+f(x-h))/h**2
    return sd


h = 10**-5
xs = 0.1
converge = 10**6
counter = 0

print("Iteration","\t","f'(x)","\t","\t","f''(x)","\t","x")
print()

while abs(converge)>10**-7:
    counter=counter+1
    xn=xs-fd(xs)/sd(xs)
    converge = xn-xs
    print(counter,"\t","\t","%0.4f" % fd(xs),"\t","%0.4f" % sd(xs),"\t","%0.4f" % xn)
    xs=xn

print()
print("Solution converged at iteration ",counter)
```

Figure 5.12 Python code for Newton's method.

Iteration	First Derivative	Second Derivative	Estimate of root
1	0.9350	–0.6341	1.5746
2	0.2811	–0.2909	2.5409
3	0.0656	–0.1648	2.9391
4	0.0076	–0.1277	2.9988
5	0.0001	–0.1227	3.0000
6	0.0000	–0.1226	3.0000
7	0.0000	–0.1226	3.0000

FIGURE 5.13 Newton's method output.

It should be noted that both these methods suffer from the problem that the solution found may be a local turning point rather than a global maximum or minimum. If there are multiple turning points for the function, then the solution found by these algorithms will be sensitive to the initial interval chosen, in the case of the bracketing method, or the initial guess for the solution, in the case of Newton's method. An additional problem in the case of Newton's method is that it will fail if the function has the property that $f''(x_k)=0$ for any x_k encountered as part of the search process. Having said that, Newton's method generally provides a very efficient, and robust, method for finding turning points in a wide variety of applications.

REVIEW EXERCISES – SECTION 5.4

1. Consider the function $f(x)=x^3/3-4x+1$. Using the initial interval $x_L=1$ and $x_U=5$, show that the bracketing method finds a local minimum in two iterations.
2. Using Newton's method, find the local maximum point of the function $f(x)=\ln x-x$ to an accuracy of two decimal places, using the starting value $x_0=0.5$.

CHAPTER 6

OPTIMIZATION OF MULTIVARIABLE FUNCTIONS

6.1 MULTIVARIABLE FUNCTIONS

> Multivariable functions allow for more than one input variable. If there are two input variables, then we can represent such functions as surfaces in three-dimensional space.

A multivariable function is a function in which several inputs are mapped to a single output variable. For example, the equation $z = x^2 + y^2$, where x and y are real numbers, can be thought of as a function that takes two input variables x and y, and produces a single output variable z. More formally, we say that this function maps the set of pairs of real numbers to the set of real numbers greater than zero. The set of pairs of real numbers (x, y) is generally written as $\mathbb{R}^2$. This generalizes so that, for functions where the input consists of n real numbers, we write the input set as $\mathbb{R}^n$.

When we considered the case of single-variable functions in the previous chapter, we found it useful to represent these geometrically. For example, in many cases, it was possible to represent a function as a curve in two-dimensional space. This, in turn, made it possible to give an intuitive explanation of many of the important results of calculus such as the nature of local maxima, minima, and points of inflexion. While it is more difficult to represent multivariable functions geometrically, we can at least do something similar when there are two input variables, that is, functions of the form $z = f(x, y)$. In these cases, we can often represent the function geometrically as a surface in three-dimensional space. By doing this, we can illustrate some of the important results of multivariable calculus, which will generalize to cases in which there are even more input variables.

Consider the function $z = 3x - 2y$ where x and y are real numbers. This function maps $\mathbb{R}^2$ to the full set of real numbers because for any real number z, we can find combinations of x and y for which $z = 3x - 2y$. Geometrically, we can think of this function as a plane in three-dimensional space, as shown in Figure 6.1. This property generalizes to all linear relationships in that any function that can be written in the form $z = ax + by + c$, where a, b, and c are parameters, will take the form of a plane. This is analogous to the case of single-variable functions where linear relationships can be represented by straight lines in the Cartesian plane.

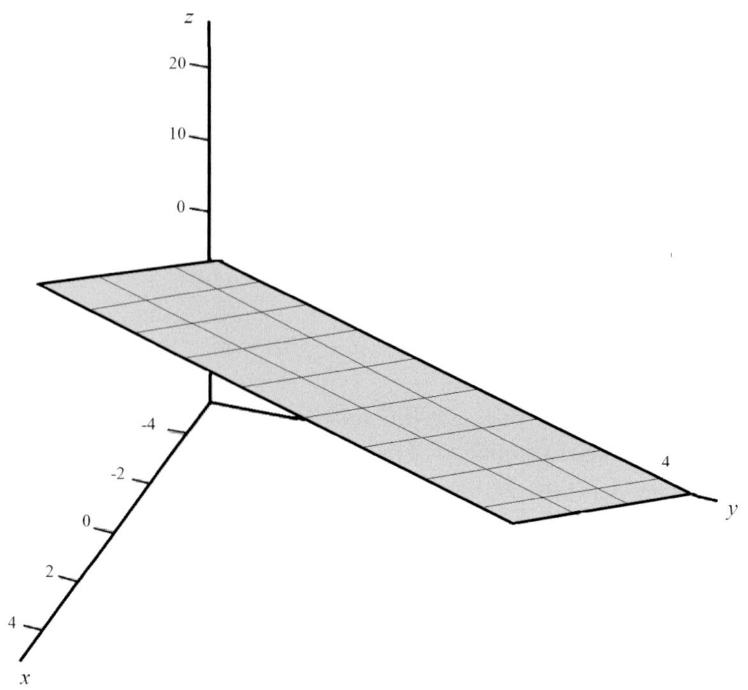

FIGURE 6.1 $z = 3x - 2y$ as a plane in three-dimensional space.

When we consider nonlinear functions, the surface representing the function will take on more complex shapes. For example, consider the function $z = 4 + 2x^2 + 3y^2$, where x and y are real numbers. This function maps $\mathbb{R}^2$ to the set of real numbers which are greater than, or equal to, four. A plot of the surface which represents this function is given in Figure 6.2. The plotted function shows a curved surface, in which there is a clear minimum point. This surface has a clear minimum point when $x = y = 0$ which gives $z = 4$. We can see that $x = y = 0$ is a minimum because, for any nonzero values of x and y, x^2 and y^2 will both be greater than zero.

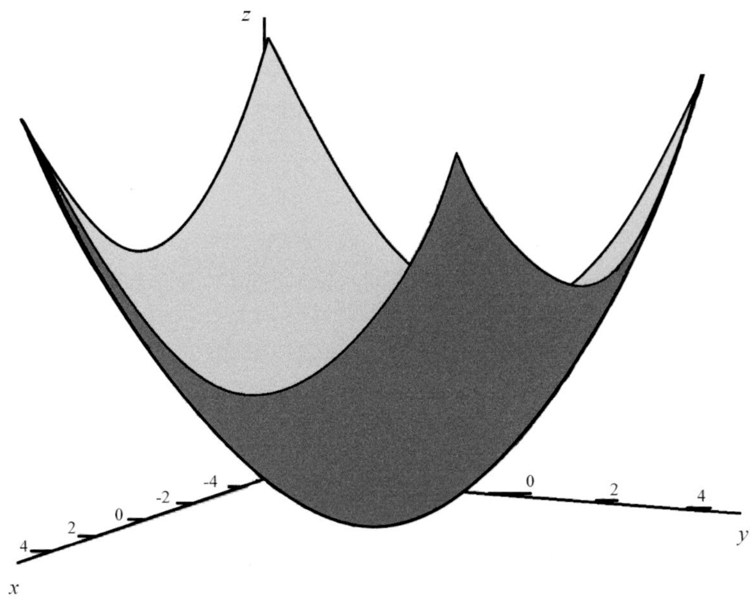

FIGURE 6.2 $z = 4 + 2x^2 + 3y^2$ as a surface in three-dimensional space.

Now, suppose we have a function of the form $z = f(x, y)$, and we fix one of the input variables while allowing the other to change. For example, let $y = \overline{y}$ and x is any real number. Setting $y = \overline{y}$ creates a function $z^* = f(x, \overline{y})$ which will map the real numbers to some subset of the real numbers. We can think of this in geometric terms as taking a cross-section of the three-dimensional surface which represents the function to create a curve in two-dimensional space. This is what we have done in Figure 6.3 which shows two "slices" of the function $z(x, y) = 4 + 2x^2 + 3y^2$ of the form $z(x, 0) = 4 + 2x^2$ and $z(0, y) = 4 + 3y^2$. In this case, the cross-sections are quadratic relationships in the (z, x) and (z, y) planes.

If a function has more than two inputs, then it becomes very difficult to represent it geometrically. However, it is still possible to apply the same mathematical tools that we will develop for functions of the form $z = f(x,y)$ to equations with three or more inputs. In such cases, we normally distinguish the inputs by writing them in the form x_i, where the i subscript represents different input variables. For example, we could define a function of the form $y = x_1^2 + x_2^2 + x_3^2$, where $x_i; i = 1,2,3$ are different real numbers. In this case the output, or y variable, will also be a real number which, in this case, will be greater than or equal to zero. A function of this kind defines a mathematical surface in four-dimensional space, but this is impossible to draw. In this chapter, we will concentrate mainly on functions with two inputs simply because this will allow us to represent them geometrically. However, all the results we derive will generalize easily to higher dimension functions.

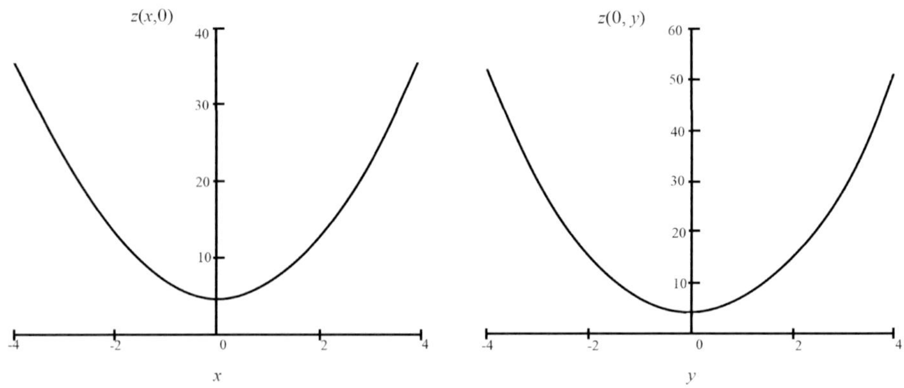

FIGURE 6.3 Cross-section planes of the function $z = 4 + 2x^2 + 3y^2$.

Some multivariable functions have the property of *homogeneity*. Homogeneity means that the function exhibits *multiplicative scaling behavior*. Consider a general function of the form $z = f(x,y)$. This function will exhibit multiplicative scaling behavior if, by increasing both the inputs by some multiplicative factor, the output increases by some power of this factor. That is, we can write $f(\lambda x, \lambda y) = \lambda^r f(x,y)$, where λ and r are real numbers. A function with this property is said to be *homogeneous of degree r*. For example, consider the linear function $z = f(x,y) = ax + by$. We have $f(\lambda x, \lambda y) = \lambda(ax + by) = \lambda z$, and therefore this function is homogeneous of degree one. Similarly, $z = f(x,y) = ax^2 + by^2$ has the property that $f(\lambda x, \lambda y) = \lambda^2(ax^2 + by^2) = \lambda^2 z$, and, therefore, this function is homogeneous of degree two. Finally, suppose we have $z = f(x,y) = ax / by$. Here, we have $f(\lambda x, \lambda y) = ax / by$, and, therefore, this function is homogeneous of degree zero. Note that not all multivariable functions have multiplicative scaling behavior.

EXAMPLE

We can show that $z(x,y) = 4x^3 + 2y^3$ is homogeneous of degree three as follows. For homogeneity, we need to find a number r such that $z(\lambda x, \lambda y) = \lambda^r z(x,y)$ for all values of λ. We have $z(\lambda x, \lambda y) = 4(\lambda x)^3 + 4(\lambda y)^3 = \lambda^3(4x^3 + 2y^3)$. Therefore, if $r = 3$, then this property is satisfied and the function is homogeneous of degree three.

EXAMPLE

We can show that the function $z(x,y) = x^2 + 2y$ is *not* homogeneous as follows. For homogeneity, we require $z(\lambda x, \lambda y) = \lambda^r z(x,y)$ for some number r for all values of λ. For this function, we need a value r which satisfies $\lambda^2 x^2 + 2\lambda y = \lambda^r x^2 + 2\lambda^r y$. Thus, we need both $\lambda^r = \lambda^2$ *and* $\lambda^r = \lambda$. This is clearly a contradiction, and the function is therefore not homogeneous.

Homogeneity, or multiplicative scaling, is often assumed for many of the functions we work with in economic and business analysis. A particularly interesting example is the *Cobb–Douglas function* which is frequently used in the analysis of production. This function takes the form

$$Y = F(K,N) = AK^{\alpha} N^{\beta},$$

where Y is output, K is capital input, and N is labor input. This function can be shown to be homogeneous as follows. We have

$$F(\lambda K, \lambda N) = \lambda^{\alpha+\beta} AK^{\alpha} N^{\beta},$$

and therefore, this function is homogeneous of degree $\alpha + \beta$. An important special case is when $\alpha + \beta = 1$. If this is the case, then the function is homogeneous of degree one, and we say that it exhibits *constant returns to scale*. This means that increasing factors of production by some proportion increases output by the same proportion. If $\alpha + \beta < 1$, then the function is homogeneous but there are diminishing returns to scale. That is, increasing capital and labor inputs in some proportion leads to a less than proportionate increase in output. Finally, if $\alpha + \beta > 1$, then the function exhibits increasing returns to scale. In this case, increasing both inputs in some proportion leads to a more than proportionate increase in output.

REVIEW EXERCISES – SECTION 6.1

1. Show that the function $z(x,y)=ax^3+by^2$, where a and b are parameters, is not a homogeneous function.
2. Show that the general quadratic function of the form $z(x,y)=ax^2+by^2+cxy$, where a, b, and c are parameters, is homogeneous of degree two.
3. Show that the Cobb–Douglas production function with constant returns to scale can be written in per capita form. That is, output per unit of labor can be written as a function of capital input per unit of labor.

6.2 PARTIAL DERIVATIVES

> Partial derivatives are calculated by allowing for small changes in one input variable while holding other variables constant. They provide the means for detecting, and identifying, turning points in multivariable functions.

Consider a function of the form $z=f(x,y)$, where x and y are real numbers. The partial derivative with respect to x gives the change in the value of the function observed when x changes, while holding the value of y constant. To distinguish this from the total derivative, which allows for changes in both variables, we use the "curly d" or "∂" notation. Thus, the partial derivative of z with respect to x is defined as

$$\frac{\partial z}{\partial x}=\text{st}\left(\frac{f(x+\Delta x,y)-f(x,y)}{\Delta x}\right).$$

In practice, the partial derivative with respect to the x variable can be obtained by differentiating the function $z=f(x,y)$ with respect to x, while treating y as constant.

EXAMPLE

Consider the function $z=3x^2+2xy+y^2$. We can calculate the partial derivative with respect to x from first principles as follows

$$\frac{\partial z}{\partial x} = \text{st}\left(\frac{3(x+\Delta x)^2 + 2(x+\Delta x)y + y^2 - 3x^2 - 2xy - y^2}{\Delta x}\right)$$
$$= \text{st}\left(\frac{6x\Delta x + 3(\Delta x)^2 + 2y\Delta x}{\Delta x}\right) = \text{st}(6x + 3\Delta x + 2y) = 6x + 2y.$$

Note that we could have obtained the same result by applying the standard power function rule for differentiation under the assumption that the variable y is constant. This is a general result, and we can easily obtain the partial derivatives of multivariable functions using the standard rules for differentiation which we developed in Chapter 4.

As with functions of one variable, there are several alternative notations for partial derivatives. For the function $z = f(x,y)$, we can use the "curly *d*" notation and write the partial derivatives with respect to x and y as $\partial z / \partial x$ and $\partial z / \partial y$. Alternatively, we can use subscript notation of the form f_x and f_y.

EXAMPLE

Consider the function $z = f(x,y) = x \ln y + e^x y$. To obtain the partial derivative with respect to x, we treat y as constant and apply the standard rules for differentiation. Similarly, to obtain the partial derivative with respect to y, we treat x as constant and differentiate with respect to y. This gives us the following results.

$$\frac{\partial z}{\partial x} = f_x = \ln y + e^x y$$
$$\frac{\partial z}{\partial y} = f_y = \frac{x}{y} + e^x.$$

For a function with a single input, the derivative gives us the slope of the tangent to the function at a particular point. We can give a similar geometric interpretation to the partial derivative as the slope of a tangent line to a cross-section of the function. Figure 6.4 shows a cross-section of the surface defined by the equation $f(x,y) = x \ln y + e^x y$, where we have fixed the x value at $x = 1$. The partial derivative function gives us the slope of tangent functions to this cross-section. In this case, at the point (1,1), the slope of the tangent line is equal to $1/1 + \exp(1) = 3.7182\ldots$

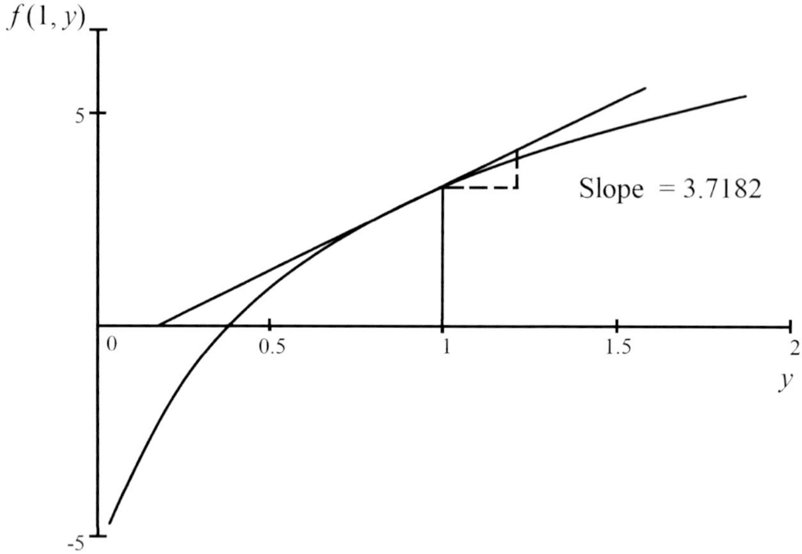

FIGURE 6.4 Tangency of a cross-section.

The partial derivatives themselves define multivariable functions in the x and y variables. We can therefore calculate higher-order partial derivatives in the same way as we did for single-variable functions by differentiating the partial derivative functions again with respect to the input variables. The notation for higher-order partial derivatives is an extension of the notation for the first-order partial derivatives. For the function $z = f(x,y)$, we write the second-order partial derivatives as $f_{xx} = \partial^2 z / \partial x^2$ and $f_{yy} = \partial^2 z / \partial y^2$.

EXAMPLE

Consider the function $z = f(x,y) = 4x^3 + 2y^2 + 3xy$, where x and y are real numbers. The first-order partial derivatives of this function are

$$f_x = \frac{\partial z}{\partial x} = 12x^2 + 3y \qquad f_y = \frac{\partial z}{\partial y} = 4y + 3x.$$

These first-order partial derivatives can be differentiated again with respect to x and y to give the second-order partial derivatives.

$$f_{xx} = \frac{\partial^2 z}{\partial x^2} = 24x \qquad f_{yy} = \frac{\partial^2 z}{\partial y^2} = 4.$$

We can also define the *cross-partial derivative* as the function obtained by first differentiating with respect to one variable and then differentiating with respect to the other. Providing that the function $z = f(x,y)$ is continuous, the order in which this process takes place will not matter.

EXAMPLE

Again, consider the function $z = f(x,y) = 4x^3 + 2y^2 + 3xy$. The cross-partial derivative can be calculated by either first differentiating with respect to x and then by y

$$\frac{\partial}{\partial y}\left(\frac{\partial z}{\partial x}\right) = \frac{\partial}{\partial y}\left(12x^2 + 3y\right) = 3$$

or by first differentiating with respect to y and then by x

$$\frac{\partial}{\partial x}\left(\frac{\partial z}{\partial y}\right) = \frac{\partial}{\partial x}\left(4y + 3x\right) = 3.$$

Partial derivatives of order three and higher are written using either subscripts or by indicating the order the curly d notation in conjunction as shown below

$$f_{\underbrace{xx\ldots x}_{n \text{ times}}} = \frac{\partial^n z}{\partial x^n}.$$

For example, the third-order partial derivatives of our example function are given by the expressions

$$f_{xxx} = \frac{\partial^3 z}{\partial x^3} = 24 \text{ and } f_{yyy} = \frac{\partial^3 z}{\partial y^3} = 0.$$

The partial derivatives for economic relationships often have a meaningful economic interpretation. For example, consider the production function $Y = F(K,N)$, where Y is output, K is capital input, and N is labor input. The first-order partial derivatives, $\partial Y / \partial K$ and $\partial Y / \partial N$ give us the marginal products of capital and labor, respectively. It is often assumed that these are positive. Similarly, the second-order partial derivatives, $\partial^2 Y / \partial K^2$ and $\partial^2 Y / \partial N^2$ give the rate at which the marginal product changes as one factor varies while holding the other constant. The assumption that there are diminishing returns to scale of capital and labor is equivalent to assuming that their respective second-order partial derivatives are negative.

EXAMPLE

A consumer derives utility from consuming two goods x_1 and x_2 according to the function $u(x_1, x_2) = \ln(x_1) + 2\ln(x_2)$ where x_1 and x_2 are always positive numbers. Show that the marginal utility of consumption is always positive for both goods and that there is diminishing marginal utility in both cases.

The marginal utilities are given by the partial derivatives of the function. These are

$$\frac{\partial u}{\partial x_1} = \frac{1}{x_1}$$
$$\frac{\partial u}{\partial x_2} = \frac{2}{x_2}.$$

Since the consumption of both goods is always positive, it follows that both the marginal utility functions are also both positive. For diminishing marginal utility, we require the second-order partial derivatives to be negative. We have

$$\frac{\partial^2 u}{\partial x_1^2} = -\frac{1}{x_1^2}$$
$$\frac{\partial^2 u}{\partial x_2^2} = -\frac{2}{x_2^2}.$$

These expressions are always negative when x_1 and x_2 are positive and therefore diminishing marginal utility is always a feature of this functional form.

REVIEW EXERCISE – SECTION 6.2

1. For the following functions, find all the first-order partial derivatives.

(a) $z = f(x, y) = \dfrac{x^3}{y}$

(b) $z = f(x, y) = x\exp(y)$

(c) $z = f(x, y) = (x^2 + y^2)^3$

2. For the function $z = 3x^2 + 4y^2 - 2x^2y$, where x and y are real numbers, find the second-order partial derivatives and the cross-partial derivatives. Show that the order of calculation for the cross-partial derivative is not important in this case.

3. Consider the Cobb–Douglas production function with constant returns to scale $Y = K^a N^{1-a}$, where $0 < a < 1$. Show that

   (a) The marginal products of capital and labor are both positive.

   (b) There are diminishing returns to both capital and labor when the other factor of production is held constant.

6.3 DIFFERENTIALS AND THE TOTAL DERIVATIVE

Suppose we have a function of the form $z = f(x,y)$ where x and y are real numbers. The *total differential* measures the overall effect on z of small changes in the input variables x and y. If the partial derivatives of the function exist, then we can write

$$dz = \frac{\partial z}{\partial x}dx + \frac{\partial z}{\partial y}dy, \tag{6.1}$$

where dz, dx, and dy are infinitesimal changes in each of the variables. The *increment* of z in response to small changes in x and y is defined as

$$\Delta z = f(x + \Delta x, y + \Delta y) - f(x,y).$$

where Δx and Δy are infinitesimal changes in x and y.

The total differential and the increment are related to each other through the *increment theorem* for the two-variable function. This is closely analogous to the increment theorem for a single-variable function and can be stated as follows

$$\Delta z = dz + \varepsilon_1 \Delta x + \varepsilon_2 \Delta y$$

where ε_1 and ε_2 are infinitesimals that depend on x, y, Δx, and Δy. This theorem's proof follows the same procedure as the increment theorem for a single variable. It is not given here because, although it is straightforward, it is also quite lengthy and distracts us from the main theme of the chapter. Instead, we will give two examples to illustrate the relationship.

EXAMPLE

Consider the function $z = 2x^2 + 3y^2$ where x and y are real numbers. The total differential for this function is $dz = 4x\,dx + 6y\,dy$ and the increment is

$$\begin{aligned}\Delta z &= 2(x+\Delta x)^2 + 3(y+\Delta y)^2 - (2x^2 + 3y^2)\\ &= 4x\,\Delta x + 6y\,\Delta y + 2(\Delta x)^2 + 3(\Delta y)^2\end{aligned}$$

since $dx = \Delta x$ and $dy = \Delta y$, we can write this in the form $\Delta z = dz + \varepsilon_1 \Delta x + \varepsilon_2 \Delta y$ where $\varepsilon_1 = 2\Delta x$ and $\varepsilon_2 = 3\Delta y$.

EXAMPLE

Consider the function $z = xy$ where x and y are real numbers. The total differential for this function is $dz = y\,dx + x\,dy$, and the increment is

$$\begin{aligned}\Delta z &= (x+\Delta x)(y+\Delta y) - xy\\ &= y\Delta x + x\Delta y + \Delta x \Delta y\end{aligned}$$

since $dx = \Delta x$ and $dy = \Delta y$, we can write this in the form $\Delta z = dz + \varepsilon_1 \Delta x + \varepsilon_2 \Delta y$ where $\varepsilon_1 = \Delta y$ and $\varepsilon_2 = 0$. (Alternatively, we could define $\varepsilon_1 = 0$ and $\varepsilon_2 = \Delta x$.)

The equation for the total differential (6.1) is closely related to that of the *tangent plane* to the function at a point (a, b). The equation for a tangent plane of a differentiable function is given by

$$z - f(a,b) = \frac{\partial z}{\partial x}(x-a) + \frac{\partial z}{\partial y}(y-b).$$

If we take values of x and y which are infinitesimally close to a and b, then we can define $x - a = dx$, $y - b = dy$, and $z - f(a,b) = dz$, which is a restatement of the equation for the total differential. This implies that the tangent plane touches the surface defined by the function in the same way as the tangent line touches the curve defined by a single-variable function. This result will prove useful when looking for maximum or minimum points of multivariable functions.

EXAMPLE

Consider the function $z = 5x + 3y^2$ where x and y are real numbers. Find the equation of the tangent plane to this function at the point $(1,1)$.

From the definition of the tangent plane, we have $z - 8 = 5(x - 1) + 6(y - 1)$, which can be expressed more neatly as $z = -3 + 5x + 6y$. If we plot this plane and the surface defined by the function, then we see that there is a point of tangency at $(1,1)$ as shown in Figure 6.5.

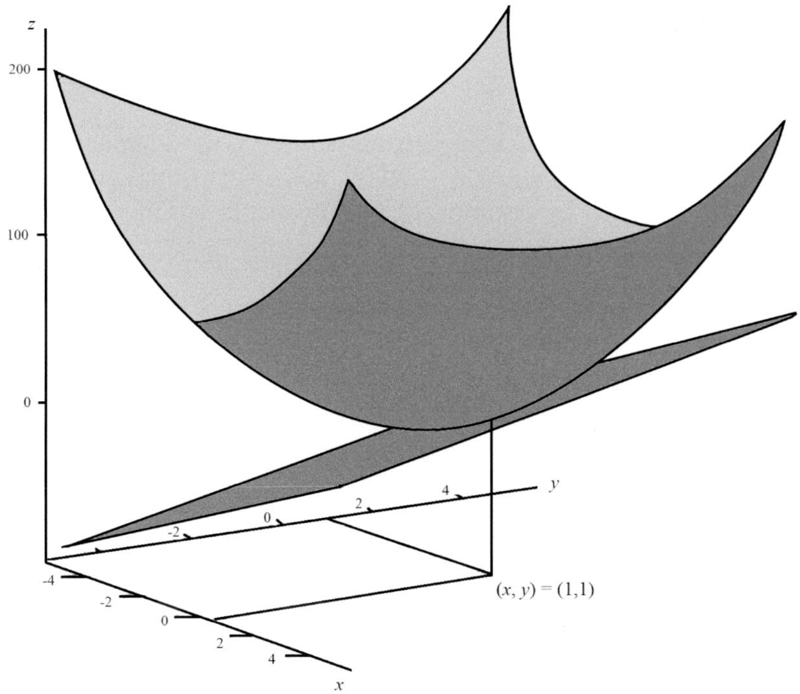

FIGURE 6.5 Plot of surface defined by $z = 5x + 3y^2$ and its tangent plane at $(x,y) = (1,1)$.

The total differential allows us to generalize the chain rule to the case of multivariable functions. Suppose we have $z = f(x,y)$ and both x and y depend on another variable t, the chain rule states that the derivative of z with respect to t is given by

$$\frac{dz}{dt} = \frac{\partial z}{\partial x}\frac{dx}{dt} + \frac{\partial z}{\partial y}\frac{dy}{dt}. \tag{6.2}$$

We can prove this using the increment theorem. From the increment theorem, we have

$$\Delta z = \frac{\partial z}{\partial x}\Delta x + \frac{\partial z}{\partial y}\Delta y + \varepsilon_1 \Delta x + \varepsilon_2 \Delta y . \tag{6.3}$$

Dividing through by Δt gives us

$$\frac{\Delta z}{\Delta t} = \frac{\partial z}{\partial x}\frac{\Delta x}{\Delta t} + \frac{\partial z}{\partial y}\frac{\Delta y}{\Delta t} + \varepsilon_1 \frac{\Delta x}{\Delta t} + \varepsilon_2 \frac{\Delta y}{\Delta t} .$$

The derivative of z with respect to t is defined as the standard part of the expression given in (6.3). Since ε_1 and ε_2 are infinitesimal and both $\Delta x / \Delta t$ and $\Delta y / \Delta t$ are finite by assumption, this proves the result given in equation (6.2).

EXAMPLE

Let $z = xy$ and let $x = x_0 e^{g_1 t}$ and $y = y_0 e^{g_2 t}$, where t is time. This assumes that the inputs of the function grow at constant proportional growth rates which are independent of each other. The chain rule gives us the following expression for the derivative of z with respect to time.

$$\frac{dz}{dt} = \left(g_1 x_0 e^{g_1 t}\right) y + \left(g_2 y_0 e^{g_2 t}\right) x = \left(g_1 + g_2\right) xy .$$

Since $z = xy$, we can divide both sides by z to obtain $\frac{1}{z}\frac{dz}{dt} = \left(g_1 + g_2\right)$.

Therefore, z also grows at a constant proportional rate equal to the sum of the growth rates of the inputs.

The total differential can also be used to find the *total derivative* of a function. The total derivative is useful when the inputs of the function are related to each other through another equation. Suppose we have $z = f\left(x, y\right)$ and $y = g\left(x\right)$. The differentials of these two equations can be written as

$$dz = \frac{\partial z}{\partial x}dx + \frac{\partial z}{\partial y}dy$$
$$dy = g'\left(x\right)dx.$$

These equations can be combined to give the single expression

$$\frac{dz}{dx} = \frac{\partial z}{\partial x} + \frac{\partial z}{\partial y} g'(x) = \frac{\partial z}{\partial x} + \frac{\partial z}{\partial y}\frac{dy}{dx}, \tag{6.4}$$

This is the *total derivative of z with respect to x*. Equation (6.4) shows that the total effect of a change x on the variable z is the sum of a direct effect, given by the partial derivative $\partial z / \partial x$, and an indirect effect produced by the effect of the change in x on the variable y, which then, in turn, affects z. The indirect effect is given by the expression $(\partial z / \partial y) dy / dx$.

EXAMPLE

Suppose we have $z = 4x^2 + \frac{1}{3}y^3$ and $y = 5x$. The total derivative of z with respect to x is equal to

$$\frac{dz}{dx} = 8x + 5y^2 = 8x + 125x^2.$$

EXAMPLE

An agent has utility function $u(c_1, c_2)$, where c_1 is consumption of good 1, and c_2 is consumption of good 2. Consumption of goods 1 and 2 are linked through the budget constraint $p_1c_1 + p_2c_2 = m$ where m is income and p_1 and p_2 are the prices of the two goods. The effect on utility of an increase in the consumption of good 1 is given by the total derivative.

$$\frac{du}{dc_1} = \frac{\partial u}{\partial c_1} - \frac{\partial u}{\partial c_2}\frac{p_1}{p_2}. \tag{6.5}$$

This equation shows that the change in utility resulting from a change in consumption of good 1 consists of two parts, the direct effect $\partial u / \partial c_1$, and the indirect effect resulting from the induced change in consumption of good 2, $-(\partial u / \partial c_2) p_1 / p_2$.

The total differential can be used to derive relationships between variables such as the *indifference curves* of consumer theory and the *isoquants* of production theory. These are essentially contours of functions of interest along which the dependent variable is held constant. First, let us consider the case of indifference curves. Consider a consumer with utility function

$u(c_1,c_2)$. The total differential of this general utility function can be written in the form

$$du = \frac{\partial u}{\partial c_1}dc_1 + \frac{\partial u}{\partial c_2}dc_2 . \tag{6.6}$$

Now, suppose we consider changes in c_1 and c_2 which are consistent with a constant level of utility. Such a relationship is referred to as an indifference curve because the agent is indifferent between such combinations of c_1 and c_2. From (6.6), and assuming $du = 0$, we have

$$\frac{dc_2}{dc_1} = -\frac{\partial u / \partial c_1}{\partial u / \partial c_2} . \tag{6.7}$$

That is, the gradient of an indifference curve is equal to minus one multiplied by the ratio of the marginal utility of consumption for good one to that of good two. This ratio is referred to as the *marginal rate of substitution* because it gives the rate at which one good can be substituted for another while leaving the total level of utility constant.

EXAMPLE

Consider the utility function $u = c_1^{\alpha} c_2^{\beta}$. From (6.7), the general expression for the slope of an indifference curve is given by

$$\frac{dc_2}{dc_1} = -\frac{\partial u / \partial c_1}{\partial u / \partial c_2} = -\left(\frac{\alpha}{\beta}\right)\frac{c_2}{c_1} .$$

Let us consider a specific example of such a utility function where the parameters α and β are both equal to one half. The indifference curves for such a function will $-(c_2 / c_1)$. Figure 6.6 shows a family of such curves drawn for different constant values of utility. Moving outwards from the origin, we set the value of u at 10, 20, and 30 to obtain the curves shown. This is termed the "indifference map." In all cases, the curves eventually approach the horizontal axis asymptotically as c_2 approaches zero and c_1 tends to infinity. This reflects the assumption of diminishing marginal utility, which is consistent with the functional form chosen. As c_1 tends to infinity, the marginal utility of consumption from this good tends to zero, leading to a flattening of the curve.

By the same reasoning, the curve approaches the vertical axis asymptotically as c_1 approaches zero and c_2 tends to infinity. This is a characteristic shape for indifference curves with the assumption of diminishing marginal utility.

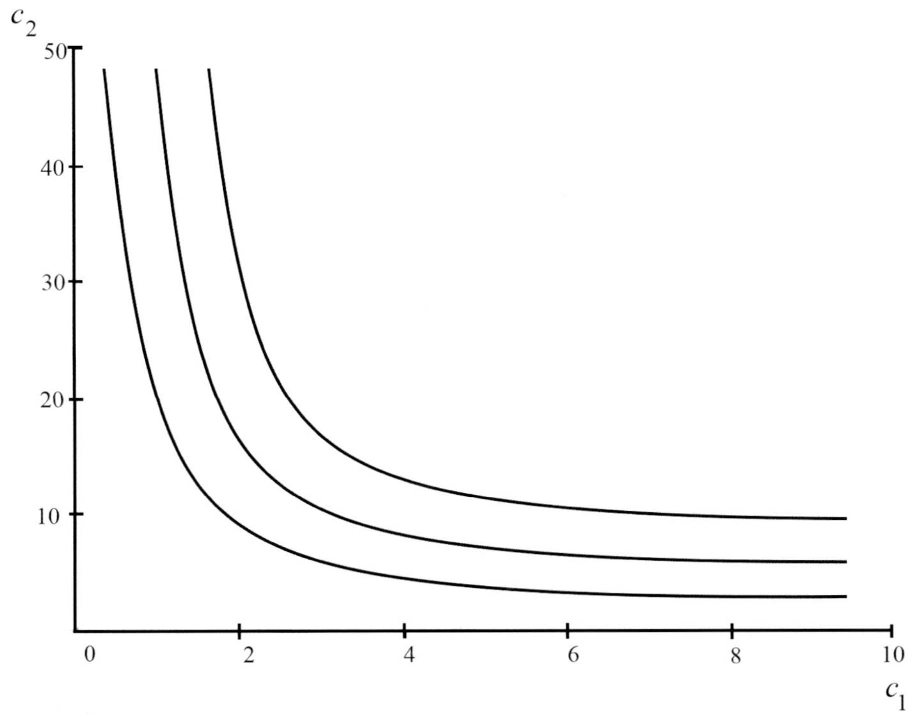

FIGURE 6.6 Indifference map for utility function $u = c_1^{0.5} c_2^{0.5}$.

A similar construction applies in the case of production theory. Consider the production function $Y = F(K, N)$, where Y is output, K is capital input, and N is labor input. The *isoquants* of this function consist of combinations of capital and labor inputs which are consistent with a fixed level of output. If the production function is differentiable in both inputs, then we can derive the slope of the isoquants using the total differential. We have

$$dY = \frac{\partial Y}{\partial K} dK + \frac{\partial Y}{\partial N} dN \cdot$$

By setting $dY = 0$, we can show that the slope of the isoquants is equal to minus one multiplied by the ratio of the marginal products of the two factors of production. That is

$$\frac{dK}{dN} = -\frac{\partial Y / \partial N}{\partial Y / \partial K}.$$

This gives us the *marginal rate of technical substitution,* which tells us that the rate at which we must increase the input of one factor as we reduce the input of another in order to maintain a constant level of output.

EXAMPLE

Consider the Cobb–Douglas production function $Y = K^{1/4}N^{3/4}$. The total differential of this function can be written as

$$dY = \left(\frac{1}{4}K^{-3/4}N^{1/4}\right)dK + \left(\frac{3}{4}K^{1/4}N^{-1/4}\right)dN.$$

Setting $dY = 0$, allows us to solve for the slope of the isoquants as $dK / dN = -3(K / N)$. It follows that their shape is essentially the same as the indifference curves we derived in our treatment of consumer theory. As $N \to \infty$, the isoquants become flat, and, as $N \to 0$, they approach the vertical axis asymptotically. However, we should note that these properties are the result of our assumption of very particular (Cobb–Douglas) production technology. In this case, there are plausible alternative technologies that will generate isoquants with different shapes.

REVIEW EXERCISES – SECTION 6.3

1. For each of the following functions, write down the total differential.

   (a) $z(x,y) = 3x^2 + 2y^3 + 4xy$

   (b) $z(x,y) = x\ln y$

   (c) $z(x,y) = e^{x-y}$

2. Let $z(x,y) = \ln\left(\frac{x}{y}\right)$ and $x = A_1 \exp(a_1 t)$, $y = A_2 \exp(a_2 t)$. Using the method of total differentiation, find dz/dt.

3. A household has utility function $u(c_1, c_2) = \ln(c_1) + \beta \ln(c_2)$. Using the method of total differentiation, find the slope of the indifference curves for this function, and use your results to sketch the indifference map.

6.4 OPTIMIZATION WITH MULTIVARIABLE FUNCTIONS

In this section, we look at how to find and identify maximum and minimum points of multivariable functions using the first- and second-order partial derivatives.

In this section, we will mostly consider functions of the form $z = f(x,y)$ where x, y, and z are real numbers. Restricting attention to the case of the two-variable function allows for a more intuitive presentation, but the results generalize easily to functions with more variables. Another advantage of the two-variable structure is that it allows us to present geometric interpretations of problems using three-dimensional diagrams.

If the function $z = f(x,y)$ is continuous, then the extreme value theorem tells us that it will have both a maximum and a minimum value within any closed interval of its domain. A maximum occurs at a point (a,b) if $f(a,b) \geq f(x,y)$ for all values of x and y, which lie within a given interval. A minimum occurs if $f(a,b) \leq f(x,y)$ for all values of x and y, which lie within a given interval. As with the single-variable function, the existence of maximum and minimum points is not guaranteed when we consider open intervals. However, we can instead look for *supremum* or *infimum* points which have a similar interpretation to maximum and minimum points when applied to open intervals.

To find possible maximum or minimum points of a function, we can extend the critical point theorem and state that, if f is differentiable for all values of (x,y) in an interval, then, for the point (a,b) to be a turning point, one of the following two statements must be true

(1) Either $\frac{\partial f}{\partial x}(a,b) = 0$ and $\frac{\partial f}{\partial y}(a,b) = 0$

(2) or (a,b) is a boundary point.

EXAMPLE

Consider the function $z = f(x,y) = \frac{3}{2}x^2 + y^2 + 2xy - 7x - 6y$, where $-4 \le x \le 4$ and $-4 \le y \le 4$.

This function has an interior stationary point where both first partial derivatives are equal to zero. These are the first-order conditions for a local maximum or minimum. We have

$$\frac{\partial f}{\partial x} = 3x + 2y - 7 = 0$$
$$\frac{\partial f}{\partial y} = 2y + 2x - 6 = 0\ .$$

These can be solved to yield a critical point $(x,y) = (1,2)$. At this point, we have $z(1,2) = -19/2$, but we do not yet have any way of determining the nature of this point. The first-order conditions simply identify a candidate point. They do not tell us whether this is a maximum, a minimum, or neither of these.

To determine the nature of critical points identified by the first-order conditions, it is helpful to consider a geometrical interpretation of the problem. We have seen that a function $z = f(x,y)$ can be thought of as a *surface* in three-dimensional space. A point at which the first-order partial derivatives are equal to zero may be a local maximum or a local minimum on this surface. There is also a third possibility in the form of a *saddle-point*. These cases can be understood geometrically as follows. Suppose we take a cross-section of the surface by fixing the value of one input variable and varying the other. Then, a local maximum will be a maximum for all possible cross-sections. Similarly, a local minimum will exhibit a minimum point in all possible cross-sections. In the case of a saddle-point, however, some cross-sections will have a maximum while some will have a minimum. A graphical presentation may help to make this clear. Consider Figure 6.7, which shows the three possible cases which can occur when the first-order partial derivatives are equal to zero. All of these examples have a critical value at the point $x = y = 0$. Panel 6.6 (a) shows the function $z = -(x^2 + y^2)$ which has a local maximum at this point, panel 6.6 (b) shows the function. $z = x^2 + y^2$. which has a local minimum, and panel 6.6 (c) shows the function $z = x^2 - y^2$ which has a saddle-point.

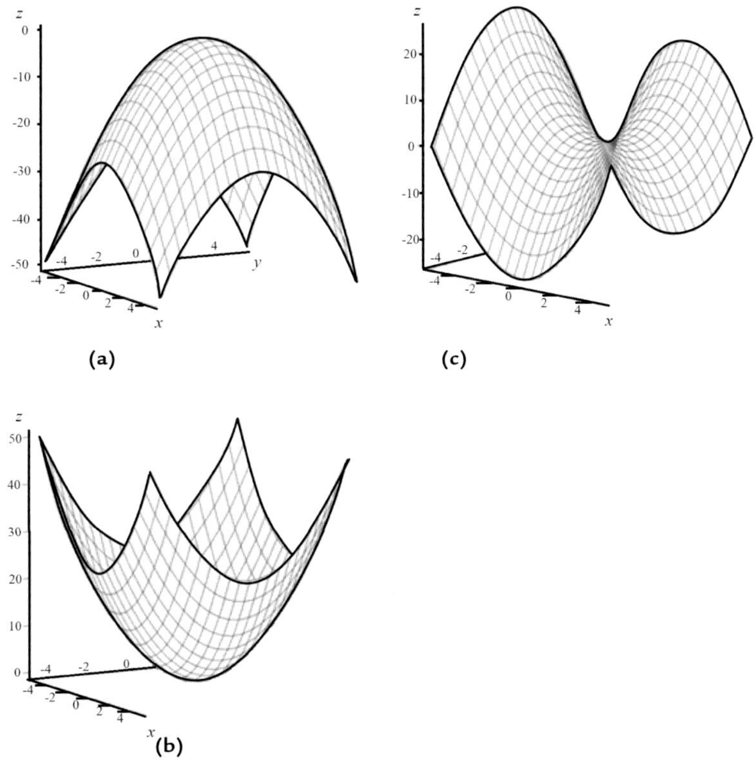

FIGURE 6.7 (a) Local Maximum, (b) Local Minimum, (c) Saddle-Point

As with the case of a single-variable function, we turn to the second-order derivatives to help us identify the nature of the critical points given by the first-order conditions. Sufficient conditions for these critical points to be points of maximum, minimum, or a saddle-point are given in (6.8)

$$\begin{array}{llll} \dfrac{\partial^2 z}{\partial x^2} < 0 & \dfrac{\partial^2 z}{\partial y^2} < 0 & \dfrac{\partial^2 z}{\partial x^2}\dfrac{\partial^2 z}{\partial y^2} - \left(\dfrac{\partial^2 z}{\partial x \partial y}\right)^2 > 0 & \text{Maximum} \\ \dfrac{\partial^2 z}{\partial x^2} > 0 & \dfrac{\partial^2 z}{\partial x^2} > 0 & \dfrac{\partial^2 z}{\partial x^2}\dfrac{\partial^2 z}{\partial y^2} - \left(\dfrac{\partial^2 z}{\partial x \partial y}\right)^2 > 0 & \text{Minimum} \\ \dfrac{\partial^2 z}{\partial x^2}\dfrac{\partial^2 z}{\partial y^2} - \left(\dfrac{\partial^2 z}{\partial x \partial y}\right)^2 < 0 & & & \text{Saddle-point} \end{array} \tag{6.8}$$

Note that these conditions are *sufficient but not necessary*. If we have

$$\frac{\partial^2 z}{\partial x^2}\frac{\partial^2 z}{\partial y^2} - \left(\frac{\partial^2 z}{\partial x \partial y}\right)^2 = 0 \tag{6.9}$$

then the second-order conditions fail to distinguish between the three possibilities. If (6.9) holds, then a critical value identified by the first-order conditions may be a local maximum, a local minimum, or a saddle-point.

The proof of the second-order conditions is not possible at this stage because it relies on properties of quadratic forms and matrices, which we have not yet covered. However, we can give some intuition regarding their roles. For a maximum, we require that both second-order partial derivatives be negative. This essentially requires that the stationary point be a maximum for all cross-sections of the surface formed by fixing either x or y. Similarly, for a minimum, we require both second-order partial derivatives to be positive, which means that the function must reach a minimum for all cross-sections. A saddle-point occurs when a critical point is a maximum for some cross-sections and a minimum in others.

EXAMPLE

Consider the function $z = f(x,y) = 3x^2 + 2x + 4y^2 - 2xy$ where the domain for both x and y is the set of real numbers. Find and identify any interior stationary points.

The first stage is to find the partial derivatives and set these equal to zero to identify critical points. This yields a pair of linear simultaneous equations in x and y.

$$\frac{\partial z}{\partial x} = 6x + 2 - 2y = 0$$
$$\frac{\partial z}{\partial y} = 8y - 2x = 0.$$

Since these are linear simultaneous equations, there is a unique solution which is given by $(x,y) = (-4/11, -1/11)$. Turning to the second-order conditions to identify the nature of the stationary point, we have

$$\frac{\partial^2 z}{\partial x^2} = 6 > 0$$
$$\frac{\partial^2 z}{\partial y^2} = 8 > 0$$
$$\frac{\partial^2 z}{\partial x^2}\frac{\partial^2 z}{\partial y^2} - \left(\frac{\partial^2 z}{\partial x \partial y}\right)^2 = 44 > 0 .$$

This satisfies the second-order conditions for a local minimum. The value of the function at this point is $z = -0.364$. Note that because the domain of the function is not a closed region, we cannot evaluate this function at its endpoints.

EXAMPLE

Consider the function $z = x^2 + 4xy + y^2$ where $-1 \le x \le 1$ and $-1 \le y \le 1$. Find any interior stationary points and find the global maximum and minimum points.

The first-order conditions can be used to identify interior stationary points. We have

$$\frac{\partial z}{\partial x} = 2x + 4y = 0$$
$$\frac{\partial z}{\partial y} = 2y + 4x = 0 .$$

These equations have solution $x = y = 0$. To determine the nature of the stationary point, we use the second-order conditions. We have

$$\frac{\partial^2 z}{\partial x^2} = 2$$
$$\frac{\partial^2 z}{\partial y^2} = 2$$
$$\frac{\partial^2 z}{\partial x^2}\frac{\partial^2 z}{\partial y^2} - \left(\frac{\partial^2 z}{\partial x \partial y}\right)^2 = -12 .$$

These conditions are sufficient to identify this point as a saddle-point. At this point we have $z = 0$.

To find the global maximum and minimum points, we must evaluate the function at the boundary points of its domain. We have $z(1,1) = 6$, $z(1,-1) = -2$, $z(-1,1) = -2$, and $z(-1,-1) = 6$. It follows that the global maximum value of the function is 6, which occurs when either $(x,y) = (1,1)$ or $(x,y) = (-1,-1)$, and the global minimum value is -2, which occurs when either $(x,y) = (1,-1)$ or $(x,y) = (-1,1)$.

REVIEW EXERCISES – SECTION 6.4

1. Consider the function $z(x,y) = x^2 + 2y^2 + 2x - 4xy$ where x and y are real numbers. Find and identify all interior stationary points for this function.
2. Consider a firm that sells in two different markets in which it faces demand curves $p_1 = 120 - q_1$ and $p_2 = 200 - 2q_2$. The cost of production is given by $C = (q_1 + q_2)^2$. Find the profit-maximizing levels of q_1 and q_2.

6.5 OPTIMIZATION WITH CONSTRAINTS

Optimization subject to constraints uses the method of Lagrange Multipliers. This introduces the idea of the shadow price of constraints which has a natural interpretation in economic theory.

To understand the impact of constraints on the optimization problem, we will first return to the single-variable problem. This may appear to be a backward step, but it allows us to introduce the idea of the *shadow price* of a constraint in an intuitive way. Shadow prices have a natural interpretation in economic theory which will prove useful for the multivariable case.

Consider an agent looking to maximize a function of the form $y = f(x)$. Now, suppose we place a restriction on the values that the variable x can take of the form $g(x) = c$, where g is a differentiable function and c is a constant. Unless $g(x) = c$ is consistent with $f'(x) = 0$, the constraint means that the first-order condition is no longer relevant. Instead, it is the constraint that determines the choice of x rather than the objective function. In such circumstances, the constraint is said to *bite*. Although the objective function no longer determines the choice of the variable, we can still use it to determine the cost of the constraint. We will now do this formally and show how this leads to the method of Lagrange Multipliers.

Our first step is to calculate the differentials of the objective function and the constraint. These can be written $dy = f'(x)dx$ and $dc = g'(x)dx$ and can be combined to give the following expression

$$dy = \frac{f'(x)}{g'(x)}dc, \tag{6.10}$$

This expression gives the cost to the agent of a marginal change in the constraint. Rearranging this expression and evaluating it at the point c gives us the *shadow price* of the constraint. That is,

$$\frac{dy}{dc} = -\frac{f'(x)}{g'(x)}. \tag{6.11}$$

This tells us how much an agent would be willing to pay for a marginal relaxation of the constraint.

EXAMPLE

Suppose we wish to find the maximum value of the function $y = \exp(x)$ subject to the constraint $x^2 = 4$. From the constraint, there are only two possible solutions $x = 2$ or $x = -2$. Since $\exp(2) > \exp(-2)$, we conclude that the maximum value of the function, given the constraint, is $\exp(2) = 7.289$. At $x = 2$, we have

$$\frac{dy}{dc} = \frac{f'(2)}{g'(2)} = \frac{\exp(2)}{4} = 1.8473.$$

The shadow price can also be shown to be generated naturally when we set up the *Lagrangian* function as part of the solution of constrained optimization problems. Let us first define a new function $L(x,\lambda)$ as shown in equation (6.12).

$$L(x,\lambda) = f(x) - \lambda(g(x) - c). \tag{6.12}$$

Equation (6.12) introduces a new variable λ which we will call the *Lagrange multiplier*. Setting the first-order partial derivatives of this function equal to zero gives

$$\frac{\partial L}{\partial x} = f'(x) - \lambda g'(x) = 0$$
$$\frac{\partial L}{\partial \lambda} = g(x) - c = 0 \, .$$

Solving the first of these equations for λ gives us $\lambda = f'(x)/g'(x)$. It follows that the Lagrange multiplier is equal to the shadow price which we derived earlier using differentials. The second equation gives us $g(x) = c$ which allows us to solve for the value of x which is consistent with the constraint.

EXAMPLE

Suppose a consumer has utility function $u(c) = \sqrt{c}$ where c is the level of consumption. Note that there is no solution to an unconstrained problem here because $u'(c) > 0$. Given this utility function, any constraint on the level of consumption will bite. Now suppose that the amount of the consumption good available to the consumer is fixed at $c = 100$. The Lagrangian function for this problem is

$$L(c, \lambda) = \sqrt{c} - \lambda (c - 100) \, .$$

Setting the first-order partial derivatives equal to zero gives

$$\frac{1}{2\sqrt{c}} - \lambda = 0$$
$$c - 100 = 0 \, .$$

Therefore, the maximum utility the consumer can achieve is $u(100) = \sqrt{100} = 10$ and the shadow price of the constraint is $\lambda = 1/\left(2\sqrt{100}\right) = 1/20$.

So far, we have restricted our attention to the single-variable problem. The Lagrangian method, however, extends easily to multivariable functions. Suppose we wish to find the critical points of a function of the form $z = f(x, y)$ where x and y are real numbers. However, there is a constraint on the values of x and y, which takes the form $g(x,y)=c$, where c is a constant. We can define the Lagrangian function for this problem as

$$L(x,y,\lambda)=f(x,y)-\lambda\left(g(x,y)-c\right). \tag{6.13}$$

If we find the partial derivatives of this function, and set them equal to zero, then we obtain the equations shown in (6.14)

$$\begin{aligned} \frac{\partial L}{\partial x} &= f_x-\lambda g_x=0 \\ \frac{\partial L}{\partial y} &= f_y-\lambda g_y=0 \\ \frac{\partial L}{\partial \lambda} &= g(x,y)-c=0. \end{aligned} \tag{6.14}$$

From the first two equations, we have

$$\lambda=\frac{f_x}{g_x}=\frac{f_y}{g_y}. \tag{6.15}$$

As with the single-variable problem, we can interpret the Lagrange multiplier λ as the shadow price of the constraint. We can solve this equation in conjunction with the constraint $g(x,y)=c$. This will give us critical values of x and y for the objective function $f(x,y)$ subject to the constraint. It will also allow us to solve for the shadow price of the constraint in the form of the Lagrange multiplier.

EXAMPLE

Consider the function $z=2x^2+3y^2+xy+x+2y$ and the constraint $x+2y=4$ where x and y are real numbers.

(a) Find the critical points of the function subject to the constraint.

(b) Find the shadow price of the constraint at the minimum.

The Lagrangian function for this problem can be written

$$L(x,y,\lambda)=2x^2+3y^2+xy+x+2y-\lambda\left(x+2y-4\right)\cdot$$

This has first-order conditions

$$\frac{\partial L}{\partial x} = 4x + y + 1 - \lambda = 0$$
$$\frac{\partial L}{\partial y} = x + 6y + 2 - 2\lambda = 0$$
$$\frac{\partial L}{\partial \lambda} = x + 2y - 4 = 0.$$

We, therefore, have a system of three linear equations in three unknown variables. The values of x, y, and λ which are consistent with these equations, are $x = 8/9$, $y = 14/9$, and $\lambda = 55/9$. These are the critical values of x and y and the value of the shadow price at the constraint.

In the example we have just considered, we identified a critical point for the problem, but we have no systematic method for determining the nature of this point. Although it is possible to find second-order conditions for Lagrangian problems, these require matrix algebra, and we have not yet covered the necessary mathematics. However, there are alternatives available to us that do require matrix methods. The first, and most direct, method is to simply evaluate the objective function for values of x and y close to the solution that are consistent with the constraint. If the value of the objective function increases when we move away from the solution, then it will be a minimum. If it falls, then the solution will be a maximum. For our example, the value of z when $x = 8/9$ and $y = 14/9$ is $128/9 = 14.22$. Now, suppose we increase the value of x slightly to 1 and reduce the value of y to $3/2$. (You might like to check that this is still consistent with the constraint). Calculating the value of the objective function for these values of x and y, gives us $z(1,3/2) = 57/4 = 14.25$. This has increased slightly, which means that the critical value we have identified is a minimum.

Another possible way to determine if critical points correspond to a maximum or a minimum is to rely on the properties of the objective function and the constraint. To do this, we will introduce the *contour plots* of the function and the constraint. A contour plot is a visual device which can be used to represent a three-dimensional surface in a two-dimensional plane. Consider the surface defined by the function $z = f(x,y)$. The contour plot of this function is constructed by fixing the value of z and then drawing the curve defined by the values of x and y which are consistent with this value. Using this method, we construct a family of curves corresponding to different values of z. An example of a contour plot for the equation $z(x,y) = xy$ is given in Figure 6.8, where we draw contours for values of z equal to 1, 2, 3, 4, and 5. Note that this device is very familiar in economics, where it is used in a variety of applications such as the *indifference map,* which is often used as a teaching device for consumer theory.

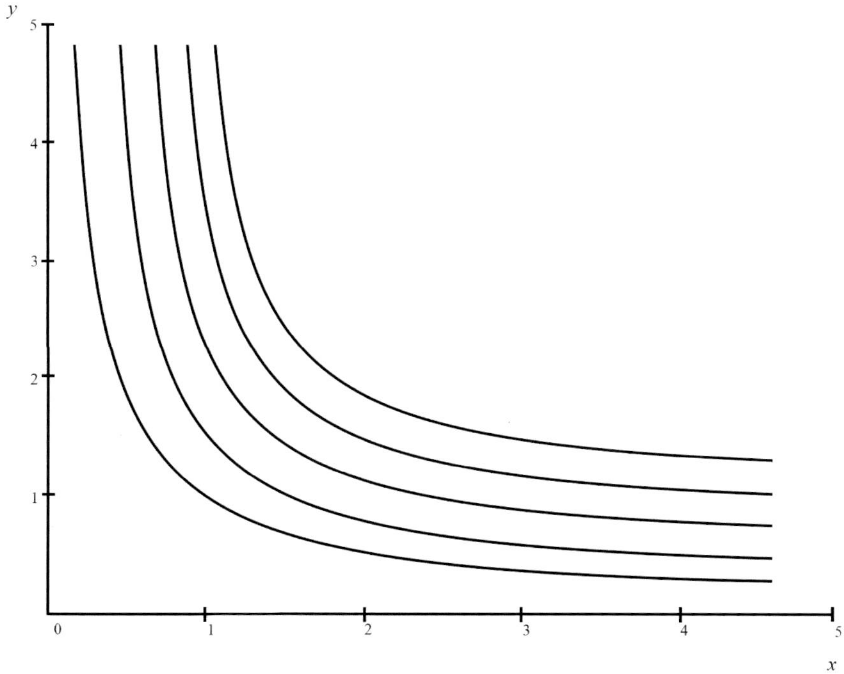

FIGURE 6.8 Contour plot for $z(x,y) = xy$ for $z = 1,2,\ldots,5$.

Contour plots are useful for understanding how the Lagrangian method identifies a critical point and in determining the nature of the point identified. Let us return to the first-order conditions for the Lagrangian problem. Rearranging (6.15), we have

$$\frac{f_x}{f_y} = \frac{g_x}{g_y}.$$

The left-hand side of this equation is the slope of a contour line for the objective function $z = f(x,y)$, and the right-hand side is the slope of contour line for the constraint $g(x,y) = c$. The Lagrangian method identifies as critical points any combinations of x and y at which the contours of the objective function are tangent to the constraint.

EXAMPLE

Suppose we wish to maximize the function $z(x,y)=xy$, where x and y are positive real numbers, subject to the constraint $0.5x+0.5y=1$.

The contours of the function $z=xy$ are curves of the form $y=\bar{z}/x$ where $\bar{z}$ is a fixed number. The constraint is a straight line that takes the form $y=2-x$ and there is a tangency point between the constraint and a contour at $(x,y)=(1,1)$ as illustrated in Figure 6.9. This is the critical point identified by the Lagrangian method.

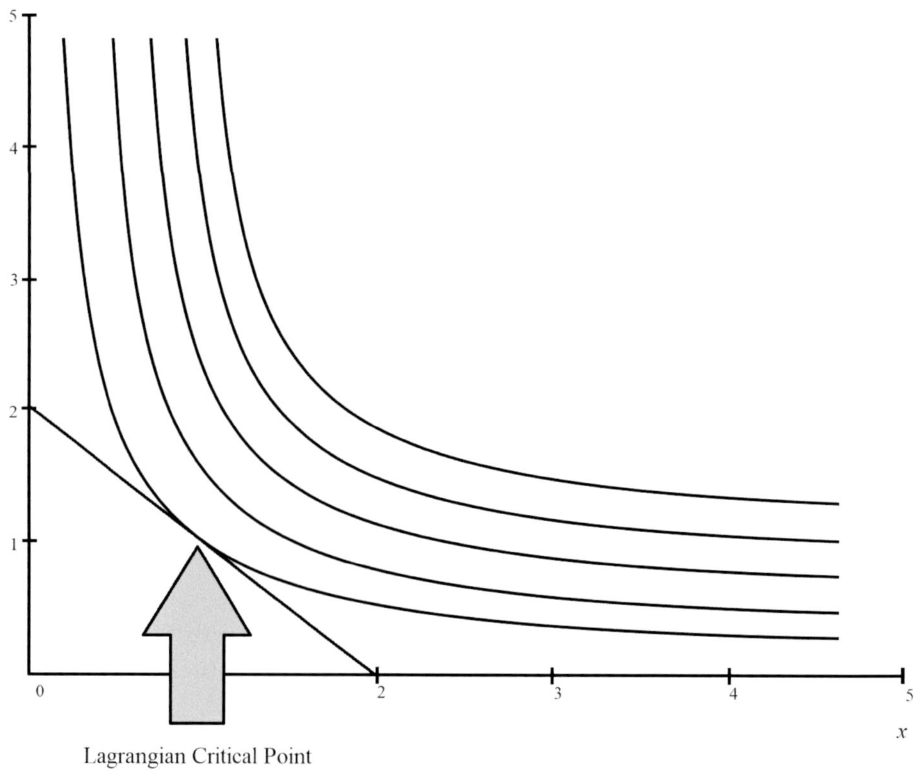

FIGURE 6.9 Determination of Lagrangian critical point.

As well as illustrating the determination of the critical point, Figure 6.9 also suggests a method by which we can identify its nature. The contours of the objective function here are strictly convex curves. By virtue of this property, any straight line drawn between points on the contour at which there is a tangency will imply a higher value of z, but these are not achievable while remaining on the constraint. It follows that the tangency identifies the contour corresponding to the highest achievable value of z, and this is, therefore, a maximum point.

The argument for determining the nature of the critical point in the Lagrangian problem generalizes quite easily. We can often use properties of the objective function and the constraint equation to determine whether the Lagrangian critical value is a maximum or a minimum. The rules for this are set out below:

For the objective function $z(x,y)$ and the constraint $g(x,y)=c$, the first-order conditions for the Lagrangian function $L(x,y,\lambda)=z(x,y)-\lambda(g(x,y)-c)$ identify:

1. A maximum if the contours of $z(x,y)$ are strictly convex and those of $g(x,y)=c$ are weakly concave.
2. A minimum if the contours of $z(x,y)$ are strictly concave and those of $g(x,y)=c$ are weakly convex.
3. A maximum if the contours of $z(x,y)$ are weakly convex and those of $g(x,y)=c$ are strictly concave.
4. A minimum if the contours of $z(x,y)$ are weakly concave and those of $g(x,y)=c$ are strictly convex.

In our example with $z(x,y)=xy$ when x and y are both positive, the contours of z are strictly convex. We can demonstrate this easily because $y=\overline{z}/x$ we have $dy/dx=-\overline{z}/x^2<0$ and $d^2y/dx^2=2\overline{z}/x^3>0$ for $x>0$. This is sufficient to ensure that the first-order Lagrangian conditions identify a maximum point.

EXAMPLE

For the function $z(x,y)=2x+3y$ where $x>0$ and $y>0$ and the constraint $3y+x^2=4$, find the first-order condition from the Lagrangian equation and determine if this corresponds to a maximum or a minimum point.

The Lagrangian function takes the form $L(x,y,\lambda)=2x+3y-\lambda(3y+x^2-4)$. The first-order conditions are

$$\frac{\partial L}{\partial x} = 2 - 2\lambda x = 0$$
$$\frac{\partial L}{\partial y} = 3 - \lambda 3 = 0$$
$$\frac{\partial L}{\partial \lambda} = 3y + x^2 - 4 = 0.$$

From the second condition, we have $\lambda = 1$ and substituting this into the first condition gives $x = 1$. We can then solve for y from the third condition to obtain $y = 1$. Therefore $(x, y) = (1, 1)$ is a critical value, but is this a maximum or a minimum? To determine this, we write the constraint as $y = 4/3 - x^2/3$. We have $dy/dx = -2x/3$ and $d^2y/dx^2 = -2/3$. The fact that both the first and second derivatives of this equation are negative is sufficient to ensure that the constraint equation is strictly concave. The combination of a strictly concave constraint and a weakly convex (straight line) objective function is sufficient to establish that this solution is a minimum point.

EXAMPLE

A consumer has utility function $u(c_1, c_2) = c_1^{1/2} c_c^{1/2}$ where c_1 and c_2 are consumption of goods 1 and 2, respectively, and $c_1, c_2 > 0$. The budget constraint is $p_1 c_1 + p_2 c_2 = m$ where p_1 and p_2 are the prices of goods 1 and 2, and m is income. Using the Lagrangian approach shows that the utility maximizing solution means that the consumer will divide expenditure equally between the goods and confirm that this is a maximum by checking that the indifference curves are strictly convex.

The Lagrangian function for this problem takes the form

$$L(c_1, c_2, \lambda) = c_1^{1/2} c_2^{1/2} - \lambda (p_1 c_1 + p_2 c_2 - m)$$

and this yields the following first-order conditions

$$\frac{\partial L}{\partial c_1} = \frac{1}{2} c_1^{-1/2} c_2^{1/2} - \lambda p_1 = 0$$
$$\frac{\partial L}{\partial c_2} = \frac{1}{2} c_1^{1/2} c_2^{-1/2} - \lambda p_2 = 0$$
$$\frac{\partial L}{\partial \lambda} = p_1 c_1 + p_2 c_2 - m = 0.$$

From the first two conditions, we have

$$\lambda = \frac{1}{2p_1} c_1^{-1/2} c_2^{1/2} = \frac{1}{2p_2} c_1^{1/2} c_2^{-1/2}$$

which simplifies to yield

$$\frac{c_1}{c_2} = \left(\frac{p_1}{p_2}\right)^{-1} = \left(\frac{p_2}{p_1}\right).$$

That is, the ratio of the consumption of the two goods is inversely related to the ratio of their prices. We can also rearrange this expression to yield $c_2 = p_1 c_1 / p_2$, and substituting this into the third condition gives us $p_1 c_1 = m / 2$. Therefore, spending on good 1 is half of total income.

To confirm that this solution is a maximum, we first note that the constraint can be written as $c_1 = (m - p_2 c_2) / p_1$ which is a linear expression and weakly concave. Therefore, if the contours of the utility function are strictly convex, then the problem satisfies the conditions for this to be a maximum. The slope of the utility function contours can be found by total differentiation of the equation $c_1^{1/2} c_2^{1/2} = \bar{u}$ which yields

$$\frac{dc_1}{dc_2} = -\frac{c_1}{c_2} < 0$$

and from the definition of the utility function, we have $c_1 = \bar{u}^2 / c_2$, which means that we can write $dc_1 / dc_2 = -\bar{u}^2 / c_2^2$. Using this expression, we can now calculate the second derivative of the contour to obtain

$$\frac{d^2 c_1}{dc_2^2} = \frac{2\bar{u}^2}{c_2^3} > 0 .$$

Since the first derivative of the contour function is always negative, and the second derivative is always positive, it follows that this function is strictly convex. Therefore, the critical point identified using the Lagrangian function is a maximum.

REVIEW EXERCISES – SECTION 6.5

1. Show that the contours of the function $z(x,y)=2x^2+y^2$, where x and y are positive real numbers, are concave functions in (x,y) space.
2. Find the values of x and y which minimize the function $z(x,y)=2x^2+y^2$ subject to the constraint $x+y=1$ where x and y are positive real numbers.
3. A firm has production function $Y=N^{0.5}K^{0.5}$ and its costs of production are equal to $0.5K+2N$ where K and N are inputs of labor and capital, respectively. If the firm needs to produce a level of output $Y=100$, find the inputs of labor and capital, which minimize the costs of production.

6.6 NUMERICAL METHODS

> In this section, we use numerical methods to solve multivariable optimization problems. If you are not already familiar with matrix methods, then you might find it useful to cover the material in Chapter 8 before you work through this material.

Numerical methods for multivariable functions use similar principles to those used for single-variable functions. We will need to jump ahead a little and make use of the concepts of vectors and matrices. However, the principles of the method remain the same as those used in Chapter 5. Basically, we look to numerical methods to solve the first-order conditions for the problem of interest and then check the second-order conditions to determine the nature of any solutions which we find.

A popular numerical technique for multivariable functions involves a modified version of Newton's method. We look for stationary points of the function $z=f(x,y)$ using the *gradient vector*. This is defined as

$$\nabla f=\begin{bmatrix}\partial z/dx\\ \partial z/\partial y\end{bmatrix}. \tag{6.16}$$

Similar to the case of the single-variable function, we look for a stationary point where the first-order partial derivatives are equal to zero, that is $\nabla f=0$. In addition to the first-order derivative vector, we will also make use of the matrix of second-order partial derivatives, or the *Hessian matrix*, to determine the nature of the solution. The Hessian matrix is defined as

$$H = \begin{bmatrix} \partial^2 z / \partial x^2 & \partial^2 z / \partial x \partial y \\ \partial^2 z / \partial x \partial y & \partial^2 z / \partial y^2 \end{bmatrix}. \tag{6.17}$$

We can calculate numerical approximations of the derivative vector and the Hessian matrix using the numerical methods for derivatives which we discussed in Chapter 4.

For a stationary point, we require $\nabla f = 0$, and we can look for candidate points by using a matrix version of Newton's method. To implement this, we start with an initial guess $\mathrm{x} = \begin{bmatrix} x^0 & y^0 \end{bmatrix}^T$, and then update this according to the formula

$$x^{k+1} = x^k - a\, H^{-1}\left(x^k\right) \nabla f\left(x^k\right) \tag{6.18}$$

where a is an adjustment parameter which is chosen to avoid diverging solutions. This expression looks forbidding, but it is just a matrix version of Newton's formula as given in equation (5.13) for a single-variable function. Once you get past the somewhat alarming notation, the method has not changed at all. We start with an initial guess for the solution and update it according to the formula (6.18). We continue updating until the solution has "converged," that is, until the change in the x values between iterations becomes negligible. The only novel element here is the introduction of the a parameter. This is included because sometimes the search process for the multivariable case will move away from the solution if we allow changes in the x values to be too large. Therefore, we normally set $0 < a < 1$ to avoid the algorithm diverging from the solution. The usual practice is to adjust a if we observe that the iterative process is not converging until we find a value that works.

To implement this algorithm, we will need quite a lot more Python code than was the case for the single-variable problem. This is a good opportunity to show how Python functions can be used to simplify a program when blocks of code are repeated several times during the execution of a program. In this case, we have a number of procedures that are repeated many times, and it proves useful to program these as functions that can be called on during the execution of the main program. Therefore, before we even start to program the iterative search for a solution, we will define functions to do the following:

1. Evaluate the function for given values of x and y.
2. Approximate the partial derivatives of the function at x and y.

3. Approximate the Hessian matrix at x and y.
4. Calculate the inverse of the Hessian matrix at x and y.
5. Update the values of x and y based on these calculations.

Figure 6.10 shows the Python code which defined functions to perform each of the five steps we have listed. These should be reasonably self-explanatory if you are familiar with Python coding, so we do not discuss them in detail here. The derivative and Hessian approximations are calculated using the centered difference method, where h is the size of the increment and is set at the start of the program.

```
def z(x1,y1):
    z=(x1-2)**2+4*x1*y1+3*(y1-1)**2
    return z

def grad1(z):
    grad = np.zeros(2)
    grad[0]=(z(x+h,y)-z(x-h,y))/(2*h)
    grad[1]=(z(x,y+h)-z(x,y-h))/(2*h)
    return grad

def hessian(z):
    hess= np.zeros((2,2))
    hess[0][0]=(1/h**2)*((z(x+h,y)-2*z(x,y)+z(x-h,y)))
    hess[1][1]=(1/h**2)*((z(x,y+h)-2*z(x,y)+z(x,y-h)))
    hess[0][1]=(1/(4*h**2))*(z(x+h,y+h)-z(x-h,y+h)-z(x+h,y-h)+z(x-h,y-h))
    hess[1][0]=hess[0][1]
    return hess

def hessinverse():
    det1 = hess[0][0]*hess[1][1]-hess[0][1]*hess[1][0]
    hessinv[0][0]=(1/det1)*hess[1][1]
    hessinv[1][1]=(1/det1)*hess[0][0]
    hessinv[0][1]=(-1/det1)*hess[0][1]
    hessinv[1][0]=hessinv[0][1]
    return hessinv

def svr():
    sv[0]=-1*(hessinv[0][0]*grad[0]+hessinv[0][1]*grad[1])
    sv[1]=-1*(hessinv[1][0]*grad[0]+hessinv[1][1]*grad[1])
    return sv
```

FIGURE 6.10 Python subroutines to calculate Newton updating formula.

The advantage of using functions to perform the operations shown in Figure 6.10 is that they can be called on to perform actions that are repeated many times in the operation of the program as well as simplifying the code in the main body of the program. In Figure 6.11, we show the code for the main body of the program. The most important part of this is the *While* loop which defines the iteration of the x vectors until the *norm* of the derivative function falls below a preset level. The norm is calculated as $\sqrt{\left(x^{k+1}-x^{k}\right)^{2}+\left(y^{k+1}-y^{k}\right)^{2}}$, and is a measure of how much the vector changes between iterations. If this value is sufficiently small, then the calculations are said to have converged. Here, we set the convergence criterion as 10^{-5}.

```
y = 100
stepsize = 0.9
enorm1=10**12
enorm2=(x**2+y**2)**0.5
epsilon=10**-5

hess = np.zeros((2,2))
hessinv = np.zeros((2,2))
sv = np.zeros(2)
counter = 0

while enorm1>epsilon*(1+enorm2):
    counter = counter +1
    if counter>100:
        exit()
    x1 = x
    y1 = y
    grad=grad1(z)
    hess = hessian(z)
    hessinv = hessinverse()
    sv=svr()
    x = x+stepsize*sv[0]
    y = y+stepsize*sv[1]
    enorm1=((x-x1)**2+(y-y1)**2)**0.5
    enorm2 =((x**2+y**2))**0.5
    print("%0.4f" %counter,"\t","%0.4f" %x,"\t","%0.4f" %y,"\t","%0.4f" %enorm1)

trace = hess[0][0]+hess[1][1]
det1 = hess[0][0]*hess[1][1]-hess[0][1]*hess[1][0]
print("Convergence achieved after ",counter," iterations")
print()
print("The gradient vector is equal to zero when")
print()
print("x = ","%0.4f" %x)
print("y = ","%0.4f" %y)
print()
print("The Hessian matrix is ")
print()
print("%0.4f" % hess[0][0],"\t","%0.4f" % hess[0][1])
print("%0.4f" % hess[1][0],"\t","%0.4f" % hess[1][1])
print()
print("The trace is ","%0.4f" % trace)
```

FIGURE 6.11 Python code to implement Newton's method for optimization of a function with two input variables.

The complete program is constructed by first imported the Python numerical and mathematical modules *import numpy as np* and *import math*, then setting out the predefined functions as given in Figure 6.10 and, finally, by setting out the main program as shown in Figure 6.11. The output of the program is determined by the sequence of *print* commands included in the main program loop and, at the end, for the final solution values. We will now go on to look at an example of how this code can be used in practice to solve a problem of interest.

EXAMPLE

Suppose we wish to find stationary points of the function $z = f(x,y) = (x-2)^2 + 4xy + (y-1)^2$. This is a relatively easy problem to solve using standard methods, and we can easily show that there is a saddle-point when $x = 0$ and $y = 1$. Our objective here, however, is to demonstrate how we can use Newton's algorithm to find a solution numerically. To illustrate the efficacy of this algorithm, we will use starting values $x^0 = y^0 = 100$ which are a long way from the solution. Despite this, we find that the solution converges quite rapidly, as shown in Table 6.1. The output here consists of the number of the iteration k, the value of x and y at iteration k, and the norm of the change in the derivative vector. After only eight iterations, we see that the gradient vector has effectively converged.

TABLE 6.1 Newton's method applied to multivariable function.

Iteration	*x* value	*y* value	Norm
1	10.0000	10.9000	126.6444
2	1.0000	1.9900	12.6644
3	0.1000	1.0990	1.2644
4	0.0100	1.0010	0.1266
5	0.0001	1.0001	0.0127
6	0.0000	1.0000	0.0013
7	0.0000	1.0000	0.0001
8	0.0000	1.0000	0.0000

```
Convergence achieved after iterations.
The gradient vector is equal to zero when
        x = 0.0000
        y = 1.0000
The Hessian matrix is

        2.0000      4.0000

        4.0000      6.0000

The trace is 8.000
The determinant is -4.0000
```

Following the iterative process used to find the solution, the code presents information about the final values of the solution. This consists of the values of x and y, the Hessian matrix at the solution, and the trace and determinant of the Hessian. These are included because they provide the second-order condition, which, in most cases, will allow us to determine if the critical point we have identified is a maximum, a minimum, or a point of inflexion.

The properties of the Hessian matrix can be used to determine the nature of any stationary points we have identified. For the two-variable problem, the second-order conditions can be stated as follows.

1. If $\det(H) < 0$ then we have a saddle-point.
2. If $\det(H) > 0$ and $tr(H) > 0$, then the stationary point is a local minimum.
3. If $\det(H) > 0$ and $tr(H) < 0$, then the stationary point is a local maximum.

If $\det(H) = 0$, then the second-order conditions fail to identify the nature of the point. However, if this is the case, then we cannot apply the modified Newton method anyway since it is necessary for H to be invertible for us to apply the iteration formula given in (6.18). In our example, the negative value of the determinant means that the point $(x,y) = (0,1)$ is a saddle-point.

REVIEW EXERCISES – SECTION 6.6

1. For a two-variable problem in which the objective function takes the form $z = f(x,y)$ show $tr(H) < 0$ and $\det(H) > 0$ are sufficient conditions for the point to be a local maximum where H is the Hessian matrix.
2. Using the code provided for this chapter, find and identify all stationary points of the function $z = (x-3)^4 + 4xy + 3(y-2)^2$.

CHAPTER 7

INTEGRATION

Differential calculus is concerned with finding the rate of change of a variable in response to changes in another variable. In graphical terms, we can think of the derivative as the slope of a tangent to a function at a point. *Integral calculus*, which we introduce in this chapter, also has a graphical interpretation as the process of finding the area under a curve between two points.

7.1 DEFINITE INTEGRATION

In this section, we define the *definite integral* of a function between two points. This can be interpreted as the area under the curve $f(x)$ between $x = a$ and $x = b$, where a and b are the lower and upper *limits of integration*.

Let us start with a very simple example. Suppose we have a function $f(x) = c$, where c is a constant. This is probably the simplest function we can define in that the curve is simply a horizontal straight line in the Cartesian plane. Now suppose we want to find the area A under this curve between the values $x = 1$ and $x = 2$. The interval $[1,2]$ is referred to as the *interval of integration* and the function $f(x) = c$ is the *integrand*. The area A here is simply the area of the rectangle with height c and width equal to the change in x. We have a standard formula for such areas which we can easily calculate. In this case, we have $A = f(x)\Delta x = c \times 1 = c$. Thus, the *definite integral* of the function $f(x) = c$ between the lower limit $x = 1$, and the upper limit $x = 2$, is simply equal to c.

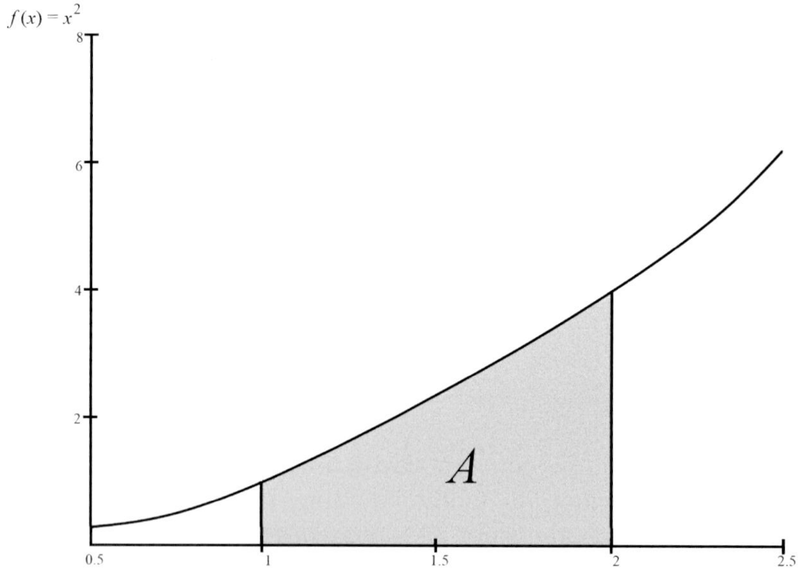

FIGURE 7.1 Area under the curve $y = x^2$.

Now let us consider a more complicated example. Suppose we wish to find the area under the curve x^2 between the values $x = 1$ and $x = 2$, as shown by the shaded area A in Figure 7.1. We do not have a standard formula for the area under curves of this type, but we can approximate it using the following procedure. First, divide the interval $x = 1$ to $x = 2$ into four subintervals, each of which has length $\Delta x = 1/4$. Next, calculate the area of each of the rectangles whose height is the value of the function at the start of the interval and whose width is the distance Δx. The approximate area under the curve is then given by the sum of the areas of these rectangles. We have

x	1	5/4	6/4	7/4	2
$f(x)$	1	25/16	36/16	49/16	4

$$A \approx \frac{1}{4} \times 1 + \frac{1}{4} \times \frac{25}{16} + \frac{1}{4} \times \frac{36}{16} + \frac{1}{4} \times \frac{49}{16} = 1.96875.$$

This process defines a *Riemann sum* for this problem, and this can be written in the form shown in equation (7.1)

$$\sum_{x=1}^{x=2} f(x)\,\Delta x. \qquad (7.1)$$

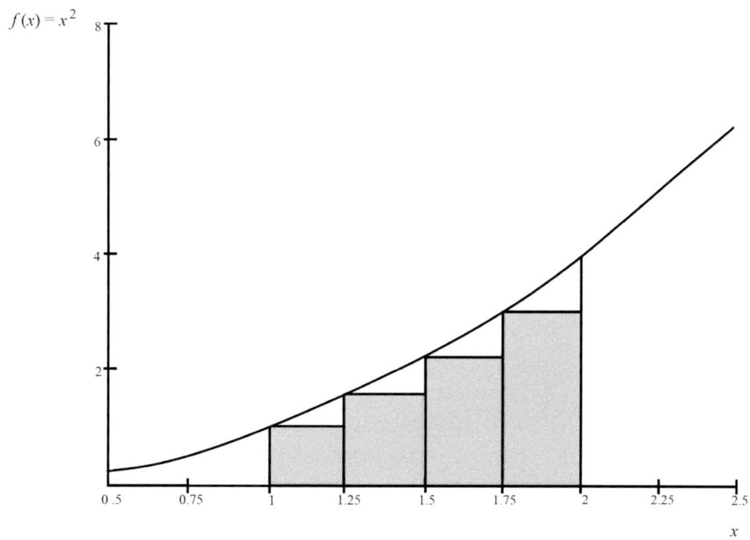

FIGURE 7.2 Approximation of the area under a curve using a Riemann sum.

Now, it is clear from Figure 7.2 that the Riemann sum we have calculated underestimates the true area under the curve. It is an underestimate because there are unshaded areas in Figure 7.2 which are under the curve that are not captured by the rectangles we have defined. However, we can improve the approximation by using a smaller interval Δx to define the Riemann sum. By taking smaller subintervals, we can eliminate part of the unshaded areas in Figure 7.2 and obtain a better approximation to the true area under the curve. This will, of course, increase the number of subintervals we use to make the calculation since the number of subintervals is equal to the total length of the interval divided by the size of interval.

Our example suggests a general approach to finding areas. Suppose we wish to find the area under the curve $y = f(x)$ between the limits $x = a$ and $x = b$ where $f(x) \geq 0$ for all points in the interval $[a,b]$. We define the Riemann sum for this general problem as

$$S(\Delta x) = \sum_{x=a}^{x=b} f(x)\Delta x. \tag{7.2}$$

That is, the Riemann sum is the sum of the rectangles whose height is the value of the function at different points in the interval $x = a$ to $x = b$ and whose width is the interval Δx, which is equal to $(b-a)/n$, where n is the number of intervals. It is obvious that increasing the number of intervals we use, or alternatively, reducing the size of the interval, will result in a better approximation of the true area under the curve.

If you have already worked through the chapter on differential calculus, you can probably see where this is going. What happens if the number of subintervals n is a positive infinite number? The answer is that the distance Δx will be a positive infinitesimal number which will be equal to the differential dx, and we can use this to define the following infinite Riemann sum

$$S = \sum_{x=a}^{x=b} f(x)dx. \tag{7.3}$$

The number S is a finite hyperreal number. It is hyperreal because it is defined as an infinite sum of infinitesimal quantities, and it is finite because $f(x)$ has maximum and minimum values B and C on the interval $[a,b]$ by virtue of the extreme value theorem. This means that the area defined by S is always less than that of the rectangle defined by multiplying the length of the interval by the maximum value of the function and is always greater than that of the rectangle defined by multiplying the length of the interval by the minimum value of the function. Thus, $C(b-a) < S < B(b-a)$ which establishes that S is bounded between two finite numbers and is, therefore, itself finite.

The definite integral of the function $f(x)$ with lower limit $x = a$, and upper limit $x = b$, can now be defined as the standard part of the Riemann sum (7.3). This is shown in equation (7.4)

$$\int_a^b f(x)dx = \mathrm{st}\left(\sum_{x=a}^{x=b} f(x)dx\right). \tag{7.4}$$

The integral sign $\int$ is used to indicate that this is an infinite sum, and the limits of integration are normally placed next to this sign, with the upper limit at the top and the lower limit at the bottom.

The definition of the definite integral as a Riemann sum lends itself to the use of numerical methods for its evaluation. For example, Figure 7.3 gives some Python code that will allow us to evaluate the definite integral of the function $y = x^2$ for any interval of integration and for any number of subintervals. Using this code, we can calculate the area under the curve $y = x^2$ between the limits $x = 1$ and $x = 2$ to a much higher degree of accuracy than given in Table 7.1. For example, if we set the number of subintervals to 10,000, we obtain the result shown in Figure 7.4. This gives the Riemann sum of 2.33318. If we compare this with the results shown in Table 7.1, then we see that the Riemann sum looks like it is converging toward the value of 7/3 as the number of subintervals increases. The proof of this will be left in the next section.

TABLE 7.1 Riemann sums for area under $y = x^2$ between $x = 1$ and $x = 2$.

Size of the subinterval Δx	Number of subintervals $n = 1/\Delta x$	Riemann sum A
1/8	8	2.148438
1/16	16	2.240234
1/32	32	2.286621
1/100	100	2.318350

```
def f(x):
    return x**2

n1 = input("Input number of subintervals ")
a1 = input("Input lower limit of integration ")
b1 = input("Input upper limit of integration ")
n = int(n1)
a = float(a1)
b = float(b1)
dx = (b-a)/n
x=a
sum1 = 0
for idx in range(1,n+1):
    sum1 = sum1+f(x)*dx
    x = x+dx

print("The Riemann sum is ",sum1)
```

FIGURE 7.3 Python code to calculate Riemann sum.

```
Input number of subintervals 10000
Input lower limit of integration 1
Input upper limit of integration 2
The Riemann sum is  2.333183334999824
```

FIGURE 7.4 Output of Python code for $a = 1$, $b = 2$, and $n = 10,000$.

REVIEW EXERCISES – SECTION 7.1

1. Evaluate the following Reimann sums
   (a) $\sum_{0}^{1} 3x^2\Delta x \qquad \Delta x = 1/4$
   (b) $\sum_{-1}^{0} 3x^2\Delta x \qquad \Delta x = 1/4$
   (c) $\sum_{0}^{1} (x-1)\Delta x \qquad \Delta x = 1/4$
   Comment on the sign of your answer to part 1(c).
2. By modifying the code given in Figure 7.3, repeat the calculations for question 1 setting the number of subintervals n at 100. Comment on the differences between these answers, and those you obtained for question 1.

7.2 THE FUNDAMENTAL THEOREM OF CALCULUS

The fundamental theorem of calculus states that we can solve for the integral of a continuous function by finding its *anti-derivative*. This makes it much easier to solve many integration problems and provides an important link between differential and integral calculus.

In the previous section, we introduced the idea of integration as a method of finding the area under a curve by use of a Riemann sum. This helps us understand the nature of integration but it would become tedious if it was necessary to do this for every problem we encountered. Fortunately, however, there are easier methods that will allow us to integrate many functions of interest. This requires us to introduce the *Fundamental Theorem of Calculus*, which provides a link between the differential calculus, which we covered in previous chapters, and the integral calculus, the current subject of interest.

We can state the fundamental theorem of calculus formally as follows. If f is a continuous function defined on a closed interval $[a,b]$ then the function

$$F(x) = \int_{a}^{x} f(u)\,du \tag{7.5}$$

is continuous and differentiable and has the property that $F'(x) = f(x)$ on the open interval (a,b) for all values of x that lie in the interval $[a,b]$. Note that the variable u acts as a *dummy variable* in this definition in that it is used in intermediate calculations but does not form part of the final result.

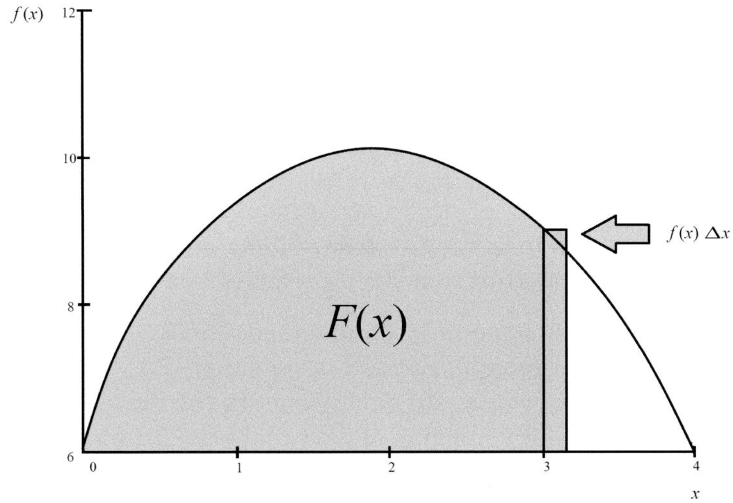

FIGURE 7.5 The integral as area under a curve.

Figure 7.5 may help provide some intuition for the fundamental theorem. Let $F(x)$ be the area under the curve $f(x)$ between the points zero and x, where in this case $x = 3$. This is the integral $\int_0^x f(u)du$. Now consider increasing the value of x by an infinitesimal amount Δx. By the increment theorem, we have

$$\Delta F(x) = f(x)\Delta x + \varepsilon\Delta x$$

where ε is infinitesimal. From this, we can write

$$\frac{\Delta F(x)}{\Delta x} = f(x) + \varepsilon$$

and, by the definition of the derivative, we have

$$\frac{dF(x)}{dx} = \text{st}\big(f(x) + \varepsilon\big) = f(x).$$

Therefore, if we wish to integrate a function $f(x)$ we should look for its *anti-derivative*. That is, we look for a function $F(x)$ which, when differentiated, gives us the original function $f(x)$.

Let us suppose we have found a function $F(x)$ which is an anti-derivative of $f(x)$. Will this be the only such function? The answer here is no, since the derivative of $F(x) + C$, where C is any constant, will also have the property that its derivative is equal to $f(x)$. For this reason, we describe the integral obtained by this method as the *indefinite integral* and write it using the following notation

$$\int f(x)dx = F(x) + C. \tag{7.6}$$

Note the absence of any limits of integration in (7.6) and the inclusion of C which is referred to as the *constant of integration*. The process of finding an anti-derivative for a function $f(x)$ is referred to as *indefinite integration*.

The *indefinite integral* given in equation (7.6) is fundamentally different from the *definite integral* defined in equation (7.4). The definite integral is a number, which gives a particular value for the area under a curve, and the indefinite integral is a function of the variable x. The indefinite integral can be used to calculate the definite integral by calculating its value at the lower and upper limits, but the two concepts are very different, and we need to keep this distinction in mind when working with them.

Finding the anti-derivative of a function is often harder than finding the derivative because there are fewer rules we can apply in this situation. In practice, the solution method often comes down to guessing an answer $F(x)$ and then confirming that it is correct by differentiating to show that we can recover the original function, that is confirming that $dF(x)/dx = f(x)$. However, there are some standard results for well-known functions, which are listed in Table 7.2.

TABLE 7.2 Anti-derivatives for standard functions.

Power function	$\int x^n dx = \frac{x^{n+1}}{n+1}$ $n \neq -1$
Reciprocal function	$\int \frac{1}{x} dx = \ln(x) + C$
Exponential function	$\int \exp(x) = \exp(x) + C$
Log function	$\int \ln(x)dx = x\ln(x) - x + C$

Some other basic rules for integrating functions are summarized in Table 7.3. These are applied when we integrate functions constructed by the combination of functions.

TABLE 7.3 Rules for indefinite integration.

Multiplication by a constant	$\int af(x)dx = a\int f(x)dx$
Sum of functions	$\int (f(x)+g(x))dx = \int f(x)dx + \int g(x)dx$
Difference of function	$\int (f(x)-g(x))dx = \int f(x)dx - \int g(x)dx$

EXAMPLE

Find the indefinite integral of the function $f(x) = 4x^3$. Using the multiplication by a constant rule and the power function rule, we have

$$\int 4x^3 dx = 4\int x^3 dx = 4\left(\frac{x^4}{4}\right) + C = x^4 + C$$

EXAMPLE

Find the indefinite integral of the function $f(x) = \ln(x) + 1$. Using the log rule and the sum rule, we have

$$\int (\ln x + 1)dx = x\ln x - x + x + C = x\ln x + C.$$

The ability to find the indefinite integral for a function simplifies the process of finding the definite integral significantly. Rather than using a Riemann sum to evaluate the area under a curve, we take the difference between the value of the indefinite integral at the upper limit and that at the lower limit. This process eliminates the constant of integration, leaving us with a single value for the definite integration problem. We can define the definite integral as follows

$$\int_a^b f(x)dx = F(b) - F(a). \tag{7.7}$$

where F is the anti-derivative of the function f. Note that the constant of integration is eliminated when we calculate the definite integral and is therefore not included in the expression given in equation (7.7). Note also that reversing the limits of integration is equivalent to multiplying the integral by minus one. We have

$$\int_b^a f(x)dx = F(a) - F(b) = -\int_a^b f(x)dx.$$

This property will prove useful when we consider the method of integrating by substitution in the next section.

EXAMPLE

Consider the function $f(x) = 1/x^2$, where x is a real number which is not equal to zero. Suppose we wish to find the area under the curve defined by this function between the limits $x = 1$ and $x = 2$.

We can write this function as $f(x) = x^{-2}$, which allows us to derive the indefinite integral as $F(x) = \int x^{-2} dx = x^{-1}/(-1) + C = -1/x + C$. To evaluate the area under the curve between the lower and upper limits, we now calculate $F(2) - F(1)$. This process can be written using the following notation

$$\int_1^2 \frac{1}{x^2} dx = \left[-\frac{1}{x} + C\right]_1^2 = \left(-\frac{1}{2} + C\right) - (-1 + C) = \frac{1}{2}.$$

Note here, the use of the square parentheses enclosing the expression for the anti-derivative, with the upper and lower limits of integration to the right outside. This is a commonly used notation when evaluating definite integrals prior to the substitution of the upper and lower limits for x. Note also that the constant of integration is always eliminated during the process of finding the definite integral and it is often omitted from the notation altogether.

EXAMPLE

Find the area under the curve $f(x) = 5\exp(x)$ between the lower limit $x = 0$ and the upper limit $x = 1$.

Using the multiplication by a constant rule and the rule for exponential functions, we have

$$\int_0^1 5e^x dx = 5\int_0^1 e^x dx = 5\left[e^x\right]_0^1 = 5(e - 1) = 8.5914.$$

So far, we have assumed that $f(x)$ is defined on a closed interval $[a, b]$. However, there are cases in which it becomes necessary to consider functions defined on an open interval. A common situation here is when the function is defined for all real numbers so that we need to consider its behavior as x approaches either ∞ or $-\infty$. Integrals evaluated using such intervals are obtained by considering the limiting value of the function as it approaches the upper or lower value and are referred to as *improper integrals*.

EXAMPLE

What is the area under the curve $f(x) = 1/x^2$ to the right of $x = 1$?

This is an improper integral because it requires evaluation for arbitrarily large values of x. However, we can evaluate this integral based on its limiting behavior. We write the problem as

$$\int_1^\infty \frac{1}{x^2}\,dx = \left[-\frac{1}{x}\right]_1^\infty .$$

Since $\lim_{x\to\infty}(-1/x) = 0$, we can ignore the upper limit, and evaluate this integral as $-(1/1) = 1$.

EXAMPLE

What is the area under the curve $f(x) = \exp(x)$ to the left of $x = 1$?

This is an improper integral because it requires evaluation of the function as $x \to -\infty$. We have

$$\int_{-\infty}^{1} e^x dx = \left[e^x\right]_{-\infty}^{1} = e - \lim_{x\to-\infty} e^x = 2.7183.$$

Note that again the value of the integral at the lower limit tends to zero as $x \to -\infty$.

Improper integrals can also arise if the function is not defined for some finite values of x and therefore has asymptotes at these values. For example, the function $f(x) = 1/(x-1)$ is not defined for $x = 1$. We will leave further consideration of such functions until we have had the chance to consider some further rules for integration in the next section.

REVIEW EXERCISES – SECTION 7.2

1. Show that the anti-derivative of $f(x) = \ln x$ is equal to $F(x) = x\ln x - x + C$ by differentiating the function $F(x)$.

2. Find the anti-derivatives for each of the following functions
   (a) $f(x) = 3x^2$
   (b) $f(x) = 2\exp(x) + x^2$
   (c) $f(x) = \ln x - \frac{1}{x^2}$
3. Evaluate the following definite integrals
   (a) $\int_0^2 (x^2 + 4x)dx$
   (b) $\int_{-1}^{1} x^3 dx$
   (c) $\int_{-\infty}^{0} \exp(x)dx$

7.3 INTEGRATION BY SUBSTITUTION AND BY PARTS

> Integration by substitution and by parts are methods for finding the indefinite integral which can simplify the problem in some cases. These rules are related to the chain and product rules for differentiation.

We have already noted that the process of integration is relatively difficult when compared to differentiation. In the case of differentiation, there are well established rules and procedures for dealing with most functions. This is not the case for integration, where the process of working back from the function of interest to the indefinite integral, or primitive function, often has to be done on a case-by-case basis. In some cases, however, we can simplify the problem by using the methods of integration by substitution and parts. In this section, we explain these methods and show how they can be applied using a few examples.

As the name suggests, integration by substitution involves using a substitution to simplify the function of interest. Consider the integral $\int (4x+1)^3 dx$. We could, in principle, expand the expression $(4x+1)^3$ and integrate this directly as a polynomial function. However, it is much easier to make use of a substitution. Let $u = 4x + 1$. We have $(4x+1)^3 = u^3$ and $dx = du/4$. By these values into the integration problem, we can write the integral as $(1/4)\int u^3 du$. This can be solved easily to give $(1/4)u^4/4 + C$ or $(1/16)(4x+1)^4 + C$. Therefore, by making an appropriate substitution, we have simplified an otherwise complicated integration problem.

EXAMPLE

Using the method of integration by substitution, calculate $\int_0^1 \left(\frac{x}{2}+4\right)^2 dx$.

This problem can be simplified by making the substitutions $u = x/2+4$ and $dx = 2du$. Making these substitutions means that the problem can be written as

$$\int_4^{9/2} 2u^2 du.$$

Note that it is important to adjust the limits of integration as well as the integrand itself if we are to calculate the indefinite integral correctly. Using this transformation gives us

$$\int_4^{9/2} 2u^2 du = 2\left[\frac{u^3}{3}\right]_4^{9/2} = \frac{2}{3}\left(\frac{729}{8}\right) - \frac{2}{3}(64) = 18.083$$

EXAMPLE

Using the method of integration by substitution, calculate $\int_0^\infty \exp(-2x)dx$.

In this case, we make the substitutions $u = -2x$ and $dx = -du/2$. Our problem now becomes

$$-\frac{1}{2}\int_0^{-\infty} e^u du = \frac{1}{2}\int_{-\infty}^{0} e^u du.$$

This is complicated since this is an improper integral in which one of the limits is $-\infty$. However, we can apply the standard methods we established earlier to write this as

$$\frac{1}{2}\int_{-\infty}^{0} e^u du = \frac{1}{2}[e^u]_{-\infty}^{0} = \frac{1}{2} - \lim_{u\to-\infty} e^u = \frac{1}{2}.$$

The method of integration by substitution can be shown to derive from the chain rule of differentiation. Suppose we wish to calculate the following integral $\int f(x)dx$. If $u = g(x)$, then we have $du = g'(x)dx$, and we can substitute for dx in the integral so that it becomes

$$\int \frac{f(x)}{g'(x)} du. \qquad (7.8)$$

Therefore, if we can choose $u = g(x)$ so that $f(x)/g'(x) = h(u)$, then the integral becomes $\int h(u)du$. The trick is to find a function $u = g(x)$ which allows us to write the integral as a function of u only and is simpler to integrate than the original function.

EXAMPLE

Find the indefinite integral $\int x\exp(x^2)dx$.

In this case we have $f(x) = x\exp(x^2)$. Suppose we choose $u = x^2$, this gives $du = 2x\,dx$. Therefore, $f(x)/g'(x)$ can be written as $\exp(u)/2$ and the integral becomes $(1/2)\int \exp(u)du$. This is considerably easier to integrate than the original problem in x. We have

$$\frac{1}{2}\int \exp(u)du = \frac{1}{2}\exp(u) + C$$

or, in terms of x

$$\int x\exp(x^2)dx = \frac{1}{2}\exp(x^2) + C.$$

EXAMPLE

Evaluate the definite integral

$$\int_0^1 \frac{4x^2}{x^3+1}dx.$$

For this problem we make the substitution $u = x^3 + 1$ which gives us $dx = du/3x^2$. Substituting these into the original problem and taking care to adjust the limits of integration, means that the problem can be written as

$$\frac{4}{3}\int_1^2 \frac{1}{u}du = \frac{4}{3}[\ln u]_1^2.$$

Since $\ln(1) = 0$, this simplifies to $\frac{4}{3}\ln(2) = 0.9242$.

Integration by parts is a useful technique when the integrand $f(x)$ is equal to the product of two functions of x. In such circumstances, we can sometimes use the product rule of differentiation to calculate the integral. Recall that the product rule states that

$$\frac{duv}{dx} = u\frac{dv}{dx} + v\frac{du}{dx} \Rightarrow u\frac{dv}{dx} = \frac{duv}{dx} - v\frac{du}{dx}$$

where u and v are functions of x. If we integrate the second form of the expression above, then we have

$$\int u\frac{dv}{dx}dx = uv - \int v\frac{du}{dx}dx. \tag{7.9}$$

This is the general expression we use for the process of integration by parts. For some integrands, this offers a simpler calculation than the original statement of the problem.

EXAMPLE

Evaluate the indefinite integral $\int \frac{\ln x}{x^2}dx$ using the method of integration by parts.

To solve this problem, we define $u = \ln x$ and $v = -1/x$. This means that $dv/dx = 1/x^2$ and we can therefore write

$$\begin{aligned}\int \frac{\ln x}{x^2}dx &= -\frac{\ln x}{x} + \int \frac{1}{x}\left(\frac{1}{x}\right)dx = -\frac{\ln x}{x} + \int \frac{1}{x^2}dx \\ &= -\frac{\ln x}{x} - \frac{1}{x} + C = -\frac{1}{x}(\ln x + 1) + C\end{aligned}.$$

EXAMPLE

Evaluate the indefinite integral $\int x\exp(-x)dx$ using the method of integration by parts.

A useful rule of thumb when applying the method of integration by parts is to allocate the choice of functions u and v so that $u(x)$ becomes simpler when differentiated. Here, for example, if we set $u = x$ and $v = \exp(-x)$, then $du/dx = 1$. We will therefore use these definitions and write the integral as

$$\begin{aligned}\int xe^{-x}dx &= -xe^{-x} - \int -e^{-x}dx \\ &= -xe^{-x} - e^{-x} + C \\ &= -e^{-x}(x+1) + C\end{aligned}.$$

EXAMPLE

Evaluate the indefinite integral $\int x\sqrt{x+1}\,dx$ using the method of integration by parts.

Again, we choose an allocation of the functions so that $u(x)$ has the simpler derivative. Therefore, we choose $u = x$ and $v = \frac{2}{3}(x+1)^{3/2}$. This allows us to write

$$\begin{aligned}\int x\sqrt{x+1}\,dx &= \frac{2}{3}x(x+1)^{3/2} - \frac{2}{3}\int (x+1)^{3/2}dx \\ &= \frac{2}{3}x(x+1)^{3/2} - \frac{4}{15}(x+1)^{5/2} + C\end{aligned}.$$

REVIEW EXERCISES – SECTION 7.3

1. Using the method of substitution, show that $\int \exp(ax)dx = \frac{1}{a}\exp(ax) + C$, where a is a real number.
2. Using the method of substitution, find $\int \frac{x}{x^2+1}dx$.
3. Using the method of integration by parts, find $\int x\exp(x)dx$.

7.4 SOME ECONOMIC APPLICATIONS

> Integration has numerous applications in economics. It is particularly useful for the conversion of streams of income and consumption over time into aggregate or "lifetime" values.

Integration is particularly useful in economics when calculating aggregates over time such as the present value of lifetime income. It provides a mathematical method for linking stocks concepts, such as wealth, with flow concepts, such as income. To do this, we must find a way of *discounting* future income streams. That is, we need to find a way of expressing future incomes in *present value* terms, where the present value of future income is the amount that an agent would be willing to pay now for a fixed amount of income at a future date. To do this, we will begin with a brief summary of the theory of compound interest.

Suppose an agent has a fixed amount of money to invest. If the annual rate of interest is equal to r, and interest is paid annually, then the initial sum a_0 will have increased in value to $a_0(1+r)$ at the end of one year. Note that we express the rate of interest in proportional terms here so that, for example, a 2% annual rate of interest corresponds to $r = 0.02$. Now suppose that, rather than being added at the end of the year, interest is added at 6-month intervals. This means that the capital sum on which the interest rate is applied is higher in the second half of the year, and therefore, the total value of the investment at the end of the year will also be higher. The value of the investment at the end of the year will now be equal to $a_0(1+r/2)^2$. In general, if interest is added n times during the year, then the value of the investment at the end of the year will be equal to $a_0(1+r/n)^n$. Interest is said to be compounded continuously in the case where n becomes arbitrarily large. Thus, the value of the investment when interest is compounded continuously is

$$a_0 \lim_{n\to\infty}\left(1+\frac{r}{n}\right)^n = a_0 \exp(r). \tag{7.10}$$

This converges on the exponential function because $\lim_{n\to\infty}(1+x/n)=\exp(x)$ is one of the ways in which the exponential function can be defined.

An illustrative example may be useful at this point. Consider an investment of \$100 at an interest rate of 10%. If the interest is added at the end of the year, then the value of the investment at that point is equal to $\$100(1+0.1)=\$110$. If it is added at six monthly intervals and compounded, then the value at the end of the year is $\$100(1+0.05)^2=\$110.25$. If it is compounded continuously, then the value at the end of the year is $\$100\exp(0.1)=\$110.52$. There may seem to be a very small difference between these values. However, one of the features of compounding processes is that apparently small differences can become quite large if they are evaluated over a long enough time period.

EXAMPLE

Suppose we invest \$100 at an annual rate of interest of 10%. If the interest is added annually, then the value at the end of a twenty-year investment period is $\$100(1+0.1)^{20}=\$672.75$. If the interest accumulates continuously, then the value of the investment at the end of 20 years is $\$100\exp(2)=\$738.91$.

In general, we can say that the value of $\$y$ after t years invested at an annual rate of interest equal to r and compounded continuously is given by the formula $\$y\exp(rt)$.

EXAMPLE

A sum of \$1,000 invested at a rate of interest of 2% for five years will yield $\$1{,}000\exp(0.02\times 5) = \$1{,}105$ (rounded to the nearest dollar.)

We can also use this relationship to calculate the *present value* of future incomes. Present values represent how much an agent values future incomes in the present. For example, suppose we have a promise of \$100 in five years. We can think of the present value as being the amount we would have to invest now to obtain this amount at the specified time. If the annual rate of interest is 5%, and it is compounded continuously, then we would need to invest a sum of $\$100\exp(-0.05\times 5) = \$77.88$ to realize such a target. The general formula for the amount necessary to obtain $\$y$ in t years, when the annual rate of interest is equal to r, is given by the formula $\$y\exp(-rt)$.

EXAMPLE

An agent knows that he will need to pay a bill of \$1,500 in two years. If the annual rate of interest is 3%. In this case, the amount he needs to invest is equal to $\$1{,}500\exp(-0.06) = \$1{,}412.65$.

We now have a method for converting future sums of money into present value terms by the method of discounting. In the examples above, we have used the rate of interest as our discount rate. However, it is possible that agents might discount the future at a different rate than the market rate of interest. More generally, we will use a rate of time discount δ which reflects the preferences of the individual. Thus, δ reflects the rate at which an agent is willing to trade current income for future income, or, alternatively, current consumption for future consumption. We generally assume that δ is positive, but it is not impossible to have a negative rate of discount if agents have a strong preference for future consumption.

By discounting future incomes, we can convert a flow variable, income, into a stock variable, wealth. Lifetime wealth is defined as the present value of the stream of income received by an agent over their entire working life. This can be calculated as the integral of the discounted present value of the agent's future income stream.

EXAMPLE

An individual has a working life of 40 years and receives \$30,000 per annum in the form of a continuous income stream. Using a discount rate of 2.5% per annum, the discounted present value of his entire income stream can be calculated using the integral shown below:

$$\int_0^{40} \$30{,}000 \exp(-0.025t)\, dt = \$758{,}545.$$

This calculation is simplified by the assumption that the income stream is constant. In practice, this will rarely be the case and income will tend to vary over the working life of the individual. More generally, we can define the lifetime wealth of the individual as:

$$\int_0^{T} y(t) \exp(-\delta t)\, dt$$

where $y(t)$ is income at date t and T is the length of the individual's working life.

EXAMPLE

An individual works for T years and has a starting salary of y_0 dollars per year. Her salary increases at a rate g throughout her working life. If future income is discounted at an annual rate given by δ, then the present value of her lifetime income is given by

$$\int_0^{T} y_0 \exp\{(g-\delta)t\}\, dt.$$

For $y_0 = \$20{,}000$, $T = 40$, $g = 0.0173$, and $\delta = 0.025$, this gives a value of \$688,532 for her lifetime wealth.

Another example of the use of integration in economic theory is the calculation of quantities such as consumer and producer surplus. Consumer surplus measures the total amount consumers would be willing to pay for a given quantity of a good over and above the actual amount they do pay. It can be measured as the area under the demand curve above the market price of the good. This is illustrated in Figure 7.6, where the shaded area illustrates the consumer surplus.

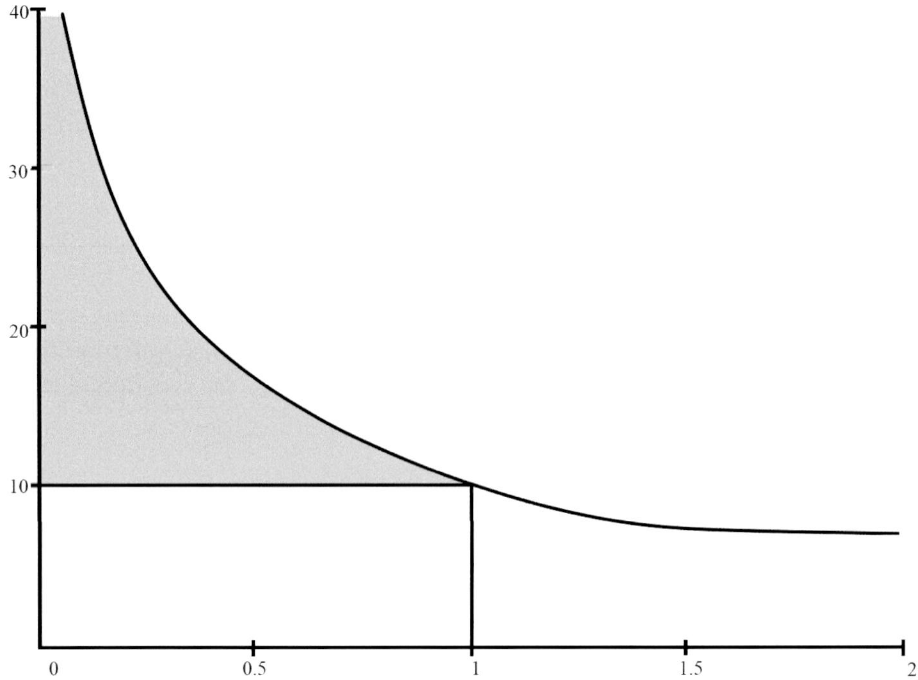

FIGURE 7.6 Consumer surplus for the demand curve $p = 10q^{-0.5}$.

EXAMPLE

Consider the demand curve $p = 10q^{-0.5}$, we can evaluate the consumer surplus as the area under the demand curve between $q = 0$ and $q = 1$ minus the amount the consumer actually pays for the product $pq = 10 \times 1 = 10$.

$$\int_0^1 10q^{-0.5}dq - 10 = \left[20\sqrt{q}\right]_0^1 - 10 = 10.$$

Note that this is an improper integral because the inverse demand function is not defined for $q = 0$. However, the area under the curve does approach a limiting value as $q \to 0$ which gives us a total consumer surplus of 10 in this case.

EXAMPLE

Consider the market demand curve $p = 100 - 10q + q^2$ where $0 \le q \le 5$. If market equilibrium price is $p = 84$, find the consumer surplus.

If $p = 84$, then, we can solve for the equilibrium quantity using the quadratic equation

$$\begin{aligned} 100 - 10q + q^2 &= 84 \\ q^2 - 10q + 16 &= 0 \end{aligned}.$$

This factorizes to give us $(q-8)(q-2)=0$ and there are, therefore, two solutions, either $q = 2$, or $q = 8$. We can ignore the second solution because it lies outside the domain of the function. Next, we can solve for the consumer surplus by integrating the function between the limits $q = 0$ and $q = 2$. This gives us

$$\int_0^2 \left(100 - 10q + q^2\right) dq = \left[100q - 5q^2 + \frac{1}{3}q^3\right]_0^2 = \frac{548}{3}.$$

This is the total area under the demand curve. To find the consumer surplus, we need to subtract the amount that consumers pay for the product which, in this case, is equal to $p \times q = 84 \times 2 = 168$. The consumer surplus is therefore equal to $\frac{548}{3} - 168 = \frac{44}{3}$.

REVIEW EXERCISES – SECTION 7.4

1. An individual has an income stream which lasts indefinitely and has initial value of \$100 but which then declines exponentially at a rate of 15% per annum. If the rate of time discount is 5% per annum, find the present value of the income stream.
2. A firm has marginal cost function $MC = 10 + 4q$, and its fixed costs are equal to 100. Find its total cost function.
3. Consider a market in which the inverse demand curve is $p = 4 - 2q$, and the market price is equal to 2. Calculate the consumer surplus associated with the market equilibrium.

7.5 NUMERICAL METHODS OF INTEGRATION

Not all integration problems have neat analytical solutions. Numerical methods provide a way of solving integrals in cases where the analytical solution is not available. In this section, we develop coding for the *trapezoidal method* and show how this can be used in practice.

The trapezoidal method provides a simple numerical algorithm for the calculation of areas under a curve. Consider the example shown in Figure 7.7, we can approximate the area under the curve between the limits $x = a$ and $x = b$ as the sum of the shaded rectangle area $(b-a)f(a)$ and the shaded triangle area $\left(\frac{b-a}{2}\right)(f(b)-f(a))$. This gives the following estimate

$$A \approx (b-a)f(a)+\left(\frac{b-a}{2}\right)(f(b)-f(a))=\left(\frac{b-a}{2}\right)(f(a)+f(b)). \quad (7.11)$$

We can think of this area as the average of two Riemann sums with interval $(b-a)$. The first, or left, Riemann sum is based on the value of the function at the lower bound $f(a)$, and the second, or right, Riemann sum is based on the value of the function at the upper bound $f(b)$.

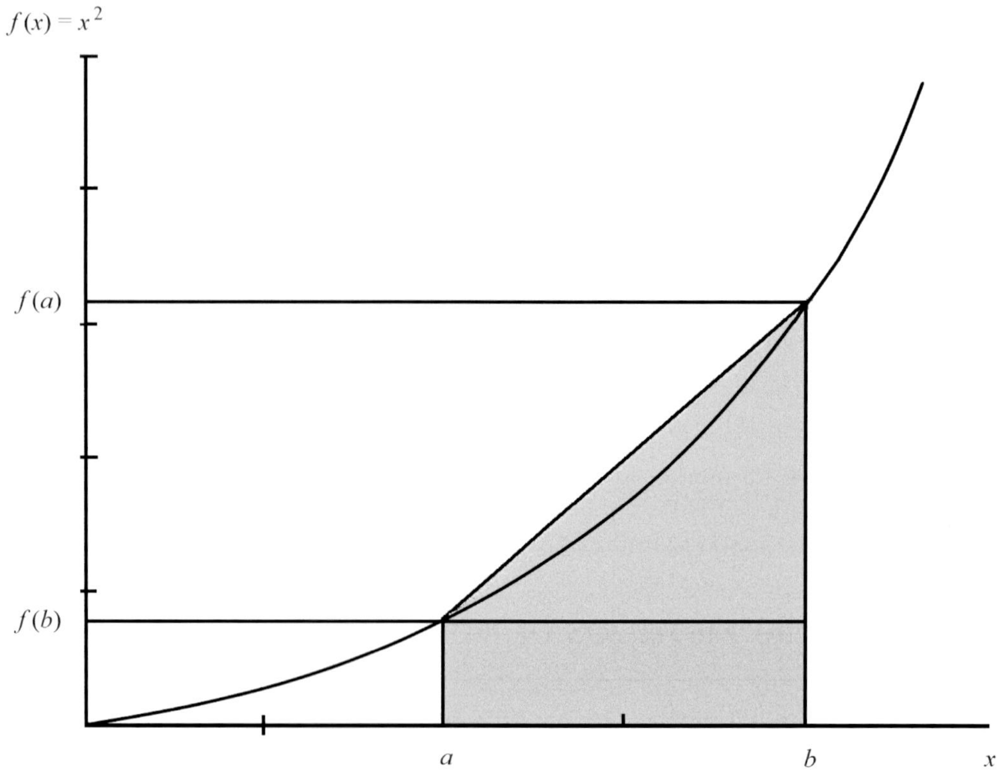

FIGURE 7.7 The Trapezoidal method.

Now suppose we divide the interval for the calculations further by taking an intermediate point $x_1 = (a+b)/2$. We now have two subintervals, the first has lower limit a and upper limit x_1, and the second has lower limit x_1 and upper limit b. Applying the same approximation to each of the subintervals and then adding them produces a new approximation for the total area which takes the form

$$A \approx \frac{hf(a)+hf(x_1)}{2} + \frac{hf(x_1)+hf(b)}{2}$$

where $h = (b-a)/2$ is the length of the subintervals. Note that $f(x_1)$ features twice in this calculation, as the upper limit of the first subinterval and as the lower limit of the second subinterval. If we increase the number of subintervals further to n, then the length of the subinterval becomes $h = (b-a)/n$, and our approximation to the area under the curve becomes

$$A \approx \frac{h}{2}\left(f(a)+2f(a+h)+2f(a+2h)+\ldots 2f(b-2h)+2f(b-h)+f(b)\right).$$

Note that all point calculations occur twice in the calculation, apart from the upper and lower limits. As n increases, the error in the calculation will be reduced, and the estimate will approach the true value of the definite integral between the lower and upper limits for x.

This method can be easily implemented using some fairly simple computer code. Figure 7.8 gives Python code for the trapezoidal method, which we can use to generate numerical estimates of definite integrals for a wide range of functions. We can also use this code to investigate how the accuracy of the estimate changes as the number of subintervals increases. To do this, we will consider an integration problem for which the analytical solution is known. This will allow us to assess how close our estimate is to the true value.

Suppose we wish to find the definite integral $\int_1^2 (1/x)dx$. We do not need to use a numerical method here because we can easily find an exact solution analytically. We have

$$\int_1^2 (1/x)dx = [\ln x]_1^2 = \ln(2)$$

and the value of $\ln(2)$ to four decimal places is 0.6931. This will give us a basis to assess the accuracy of our numerical estimates.

```
import math

def f(x):
    f = 1/x
    return f

ns = input("Input number of sub-intervals ")
n = eval(ns)
xl = 1
xu = 2
dx = (xu-xl)/n
sum1 = f(xl)
x = xl

for idx in range(1,n):
    x = x+dx
    sum1 = sum1+2*f(x)

sum1 = sum1+f(xu)
area = dx*sum1/2

print("The integral is ","%0.4f" % area," where dx = ",dx)
```

FIGURE 7.8 Python code for integration using trapezoidal method.

Now, suppose we apply the Python code given in Figure 7.8 to this problem, starting with the most basic trapezoidal estimate, which we set $n = 1$, and then increasing n to generate better estimates. The results of this process are given in Table 7.3, which shows that the error is quite large for low values of n but that the estimate converges quickly toward the true solution as we increase the number of subintervals. For $n \geq 100$, we see that the result is accurate to four decimal places.

TABLE 7.3 Calculation of the definite integral $\int_1^2 (1/x)dx$ using the trapezoidal method.

n	*Approximate Area*	*Percentage Error*
1	0.7500	8.21
2	0.7083	2.20
5	0.6956	0.37
10	0.6937	0.10
100	0.6931	7.7×10^{-3}
1,000	0.6931	6.8×10^{-3}

An alternative numerical method of integration is provided by *Simpson's rule*. This derives from a long-established approximation to integrals based on the following formula.

$$\int_a^b f(x)dx \approx \frac{b-a}{6}\left[f(a)+4f\left(\frac{a+b}{2}\right)+f(b)\right]$$

The expression on the right-hand side can be shown to be a quadratic approximation to the true integral function, which passes through the endpoints and the mid-point of the true function. This is illustrated in Figure 7.9 for the case in which $f(x)=1/x$, and the true indefinite integral function is $F(x)=\ln(x)$. The solid line shows the true function $F(x)=\ln(x)$, and the curved broken line shows the Simpson's rule, or quadratic, approximation. The simple trapezoidal estimate is shown by the broken line between the two endpoints. In this case, Simpson's rule clearly provides a more accurate approximation. As with the trapezoidal method, we can increase the accuracy of Simpson's rule by dividing the interval up into several subintervals and applying the method to each of these individually. Under the assumption that the number of subintervals n is even, this leads to the *composite Simpson's rule* formula, which takes the form.

$$\int_a^b f(x)dx \approx \frac{h}{3}\begin{bmatrix} f(a)+4f(a+h)+2f(a+2h)+\dots \\ +2f(b-2h)+4f(b-h)+f(b) \end{bmatrix}$$

where $h=(b-a)/n$ is the width of a subinterval.

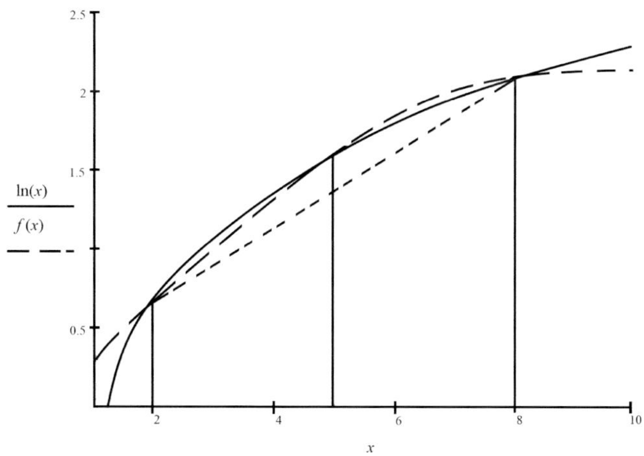

FIGURE 7.9 Simpson's rule and Trapezoidal approximations to a nonlinear function.

Table 7.4 shows the increased accuracy from using Simpson's rule rather than the trapezoidal rule. The table shows three definite integrals with known values and compares them to the estimates obtained using numerical estimators based on the trapezoidal rule and Simpson's rule with 10 subintervals in each case. The numbers in the parentheses below the estimates are the absolute values of the percentage error when the estimate is compared to the true value. In all three cases, Simpson's rule gives an answer much closer to the true value than the trapezoidal rule. Although both methods can be made more accurate by increasing the number of subintervals, Simpson's rule will always need a lower number of such intervals to achieve a given degree of accuracy.

TABLE 7.4 Comparison of trapezoidal and Simpson's rule estimates of integrals.

	Trapezoidal rule $n = 10$	**Simpson's rule** $n = 10$	**Accurate value (six decimal places)**
$\int_2^8 \ln(x)dx$	9.238031 (0.122)	9.249079 (0.002692)	9.249238
$\int_0^1 \exp(x)dx$	1.719713 (0.083)	1.718283 (0.0001164)	1.718281
$\int_0^2 x^2/(1+x^4)dx$	0.616071 (0.112)	0.616748 (0.002432)	0.616763

REVIEW EXERCISES – SECTION 7.5

Evaluate each of the following integrals by direct integration and, using the full interval, by Simpson's rule. Comment on the results.

(a) $\int_0^1 (5x+2)dx$

(b) $\int_0^1 (2x^2+3x)dx$

(c) $\int_0^1 2x^3\,dx$

(d) $\int_0^1 5x^4dx$

CHAPTER 8

MATRICES

A *matrix* is a mathematical object which consists of a rectangular array of other objects. For the purposes of this chapter, we will consider matrices in which the objects concerned are numbers. Such matrices provide a structure that allows for the simplified presentation and solution of systems of linear simultaneous equations.

8.1 MATRIX ALGEBRA

> Matrix algebra consists of a set of rules which allow us to manipulate matrix objects systematically. Using these rules, we can add, subtract, and multiply matrices to create new matrix objects.

The *elements* of a matrix are the objects contained within it which can be distinguished by their row and column numbers. For example, let us consider the object defined in

$$A = \begin{bmatrix} 1 & 4 & 5 \\ 3 & 2 & 0 \end{bmatrix}. \tag{8.1}$$

A is an example of a 2×3 matrix because it contains two rows and three columns. Each element here is associated with a particular row and a particular column. A common mathematical convention is to capitalize the symbol used to represent a matrix but to represent its individual elements using lower case notation. In our example, we write the matrix itself using the symbol *A*, but we write its individual elements as a_{ij} where i is the row number and j is the column number. Therefore, in our example, we have $a_{12} = 4$ and $a_{23} = 0$.

A *square matrix* is a matrix in which the number of rows is equal to the number of columns. For example, the matrix

$$A = \begin{bmatrix} 4 & 1 & 7 \\ 2 & 5 & -1 \\ 3 & 2 & 3 \end{bmatrix}$$

is a 3×3 square matrix. Matrices of this type have properties which are not shared with nonsquare matrices in which the number of rows and columns differ. This will become evident when we consider matrix algebra in the next section.

A *vector* is a special type of matrix which contains only one row or one column. A *row vector* is a matrix with one row but multiple columns. A *column vector* is a matrix with one column but multiple rows. These are normally written using lower case notation. For example

$$a = \begin{bmatrix} 1 \\ 2 \\ 0 \end{bmatrix}$$

is a 3×1 column vector since it has three rows and one column, while

$$b = \begin{bmatrix} 5 & 1 & 2 & 4 \end{bmatrix}$$

is a 1×4 row vector since it has one row and four columns.

Finally, we can think of individual numbers as being a special case of a matrix with one row and one column. If we are working with matrix and vector objects, then individual numbers are often referred to as *scalar* quantities to distinguish them from more general cases.

Matrices are useful because they simplify both the notation and the practice of many operations in linear algebra. For example, the solution of systems of simultaneous equations can be simplified with the application of matrix methods. To use these methods, we need to define rules for adding, subtracting, and multiplying matrices. These rules are collectively referred to as *matrix algebra*. Note that conventional algebra defined in terms of single numbers is sometimes referred to as *scalar algebra* to distinguish it from the rules which apply to matrix objects. While the rules of matrix algebra are sometimes similar to those of scalar algebra, there are important differences, and it is easy to make errors by treating matrices as if they were scalar objects.

Addition or Subtraction of Matrices

The matrix operations of addition and subtraction are closely related to the same operations for scalar objects. However, when working with matrix objects, it must be the case that the matrices concerned are *conformable*. That is, the dimensions of the matrices must be consistent. For example, if we wish to add together two matrices A and B, then they must have the same number of rows and columns. If this holds, then we either add or subtract matrices by simply adding or subtracting the individual elements.

EXAMPLE

Suppose we have matrices A and B each of which have dimensions 2×3. That is, they both have two rows and three columns.

$$A = \begin{bmatrix} 3 & 4 & 1 \\ 2 & 7 & 5 \end{bmatrix} \qquad B = \begin{bmatrix} 1 & 0 & 4 \\ 2 & 9 & 1 \end{bmatrix}.$$

Now let C be the matrix defined as $A + B$. C will also be a 2×3 matrix in which $c_{ij} = a_{ij} + b_{ij}$. Therefore, we have

$$C = A + B = \begin{bmatrix} 4 & 4 & 5 \\ 4 & 16 & 6 \end{bmatrix}.$$

Similarly, if we wish to subtract the matrix B from the matrix A, then the resulting matrix $C = A - B$ will have elements $c_{ij} = a_{ij} - b_{ij}$, which can be calculated as

$$C = A - B = \begin{bmatrix} 2 & 4 & -3 \\ 0 & -2 & 4 \end{bmatrix}.$$

The operations of matrix addition and subtraction have similar properties to their scalar equivalents. For example, matrix addition and subtraction are both associative, and therefore, the statements $(A + B) + C = A + (B + C)$

and $(A-B)-C=A-(B-C)$ remain true when A, B, and C are conformable matrices. Similarly, as in scalar algebra, addition is commutative, but subtraction is not. That is $A+B=B+A$ but, in general, $A-B\neq B-A$.

Matrix transposition

Matrix transposition is a special kind of operation that does not have a parallel in scalar algebra. Suppose A is a matrix with m rows and n columns. Its *transpose* is defined as the matrix A^T which has n rows and m columns and in which the elements of A^T are defined as $A_{ij}^T = A_{ji}$. The operation of transposing a matrix is referred to as *matrix transposition* and the superscript T is used to indicate the operation of transposition. An alternative notation is to use the 'prime' symbol to indicate transposition, that is $A' = A^T$.

EXAMPLE

Consider the matrix A, which we defined earlier. A is a 2×3 matrix, and therefore, its transpose is a 3×2 matrix. We have

$$A=\begin{bmatrix}3 & 4 & 1\\ 2 & 7 & 5\end{bmatrix}\Rightarrow A^T=\begin{bmatrix}3 & 2\\ 4 & 7\\ 1 & 5\end{bmatrix}.$$

A *symmetric matrix* is one which is unchanged by the operation of transposition, that is $A^T = A$. This can only be the case if the matrix is a square matrix.

EXAMPLE

A 2×2 matrix is symmetric if its diagonal elements are equal. That is $A^T = A$ if and only if $a_{12} = a_{21}$.

Scalar multiplication

Multiplication of a matrix by a scalar quantity simply involves the multiplication of each individual element by the same scalar quantity. Therefore, if k is a scalar, and A is a matrix, then scalar multiplication of A by k defines a new matrix C in which $c_{ij} = k a_{ij}$.

EXAMPLE

If $A = \begin{bmatrix} 3 & 1 \\ 2 & 0 \end{bmatrix}$ and $k = 2,$ then we have

$$C = kA = 2\begin{bmatrix} 3 & 1 \\ 2 & 0 \end{bmatrix} = \begin{bmatrix} 6 & 2 \\ 4 & 0 \end{bmatrix}.$$

Vector multiplication

Suppose we have a row vector a with n columns and a column vector b with n rows. We define the *scalar product* of these two vectors as the sum of the products of the individual elements. That is, we have

$$a \cdot b = \sum_{i=1}^{n} a_i b_i. \tag{8.2}$$

The term scalar product is appropriate here because, although this is an operation on vectors, the result is a single number or scalar quantity. Note that this can only be applied to conformable vectors. That is, the row vector a and the column vector b must contain the same number of elements.

EXAMPLE

Suppose we have

$$a = \begin{bmatrix} 4 & 2 \end{bmatrix} \qquad \text{and} \qquad b = \begin{bmatrix} 3 \\ 4 \end{bmatrix},$$

then the scalar product is equal to

$$a \cdot b = \begin{bmatrix} 4 & 2 \end{bmatrix}\begin{bmatrix} 3 \\ 4 \end{bmatrix} = 4 \times 3 + 2 \times 4 = 12 + 8 = 20.$$

Scalar products are also referred to as *dot products* or *inner products* of vectors. The term "dot product" derives from the notation of the dot, which is often placed between the names of the two vectors when indicating the operation. The term "inner product" is used to distinguish this definition from an alternative method for multiplying vectors known as the "outer product." We are not concerned with this method here, but you need to be aware that when the term inner product is used in this context, it simply means the scalar or dot product.

Scalar products of vectors arise frequently in economic analysis. For example, consider an economy that produces a total of n goods. Let p be the column vector whose elements are the prices of goods $i = 1,\ldots,n$ and q be the column vector whose elements are the quantities produced of each good. The total value of output is equal to the scalar quantity defined by the dot product of the transpose of vector p and the vector q. That is, we have $R = p^T q$, where R is total revenue.

Another example frequently encountered in economic statistics is where e is a vector that consists of deviations of a variable from its mean, that is, the elements of the vector e are $e_i = Y_i - \overline{Y}$, where Y_i; $i = 1,\ldots,n$ is the variable of interest. The dot product of the vector e and its own transpose defines the sum of the squared differences from the mean. That is, we have $SSD = e^T e$, where SSD is the sum of squared differences. The use of vectors in situations like this allows for much more compact notation than would be possible with scalar notation.

Matrix multiplication

Matrices A and B are conformable for the purpose of matrix multiplication if the number of columns of matrix A is equal to the number of rows of matrix B. If this is the case, then we can calculate the product of these two matrices, which we write as $C = AB$. In this case, we say that the matrix B is *premultiplied* by the matrix A, or alternatively that the matrix A is *postmultiplied* by the matrix B. This distinction is necessary because, in contrast with scalar algebra, matrix multiplication is not commutative. In general, $AB \neq BA$ and both products may not even be defined.

Suppose A is an $m \times n$ matrix and B is an $n \times p$ matrix. The matrix $C = AB$, where the matrix B is premultiplied by A, is an $m \times p$ matrix where the i, jth element is calculated by taking the scalar product of the ith row of A with the jth column of B. Thus, if $C = AB$, then we have

$$c_{ij} = \sum_{k=1}^{n} a_{ik} b_{kj}.$$

EXAMPLE 1

Let $A = \begin{bmatrix} 1 & 2 \\ 0 & 4 \end{bmatrix}$ and $B = \begin{bmatrix} 2 & 4 & 5 \\ 1 & 3 & 1 \end{bmatrix}$.

These matrices are conformable for the purposes of premultiplying matrix B by matrix A because A has two columns and B has two rows. We have

$$C = AB = \begin{bmatrix} 4 & 10 & 7 \\ 4 & 12 & 4 \end{bmatrix}.$$

Note that it is not possible to calculate the product BA because the matrix B has three columns and A only has two rows.

A visual guide may help you to understand the construction of the C matrix more clearly. Figure 8.1 shows how we calculate a typical element of the product of two matrices A and B. The matrix A is placed on the lower left and the matrix B is on the upper right. To calculate the element in row 2 column 2 of the product matrix $C = AB$ we take the vector formed by the second row of A and form the scalar product with the column vector formed by taking the second column of B. This gives us the value $c_{22} = 12$ indicated in the new matrix C shown on the lower right. Repeating this calculation for all combinations c_{ij} allows us to fill in all the elements of the product matrix.

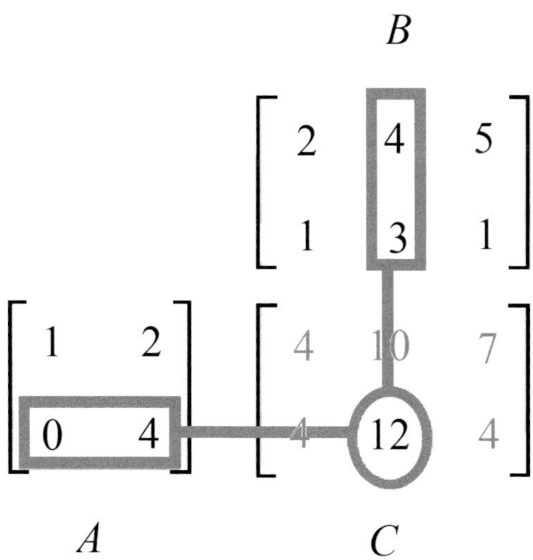

FIGURE 8.1 Premultiplication of matrix B by matrix A to create matrix C.

EXAMPLE 2

Let $A = \begin{bmatrix} 1 & 2 \\ -4 & 3 \\ 5 & 2 \end{bmatrix}$ and $B = \begin{bmatrix} 2 & -1 \\ -4 & 7 \end{bmatrix}$

These matrices are conformable for the purpose of calculating $C = AB$ because A is 3×2 and B is 2×2 , and therefore, the number of columns of A is equal to the number of rows of B. In this case, the product $C = AB$ will be a 3×2 matrix. We have

$$C = AB = \begin{bmatrix} 1 & 2 \\ -4 & 3 \\ 5 & 2 \end{bmatrix} \begin{bmatrix} 2 & -1 \\ -4 & 7 \end{bmatrix} = \begin{bmatrix} -6 & 13 \\ -20 & 25 \\ 2 & 9 \end{bmatrix}.$$

We have already noted that the operation of matrix multiplication is not commutative. Even in cases where both AB and BA are defined, it will not, in general, be true that $AB = BA$. This is easily established using a counter example.

EXAMPLE

Let $A = \begin{bmatrix} 4 & 2 \\ 3 & 4 \end{bmatrix}$ and $B = \begin{bmatrix} 1 & 2 \\ 3 & 0 \end{bmatrix}$.

We have

$$AB = \begin{bmatrix} 10 & 8 \\ 15 & 6 \end{bmatrix} \qquad BA = \begin{bmatrix} 10 & 10 \\ 12 & 6 \end{bmatrix}.$$

This example immediately establishes the result that, even if both AB and BA exist, they are generally not equivalent.

One useful property that does hold generally is that, if the matrix AB is defined, then its transpose is equal to the transpose of B postmultiplied by the transpose of A, that is $(AB)^T = B^T A^T$. This result can be very useful when expanding or simplifying complicated matrix expressions.

Proof: The result that $(AB)^T = B^T A^T$ follows directly from the definition of matrix transposition. We have $c_{ij}^T = c_{ji}$. Now c_{ji} is formed as the scalar product of the jth row of the matrix A and the ith column of the matrix B. That is,

$$c_{ji} = \sum_{k=1}^{n} a_{jk} b_{ki}.$$

We get exactly the same result if we form the scalar product of the ith row of B^T and the jth column of A^T.

EXAMPLE

Let $A = \begin{bmatrix} 4 & 1 & 2 \\ 3 & 1 & 7 \end{bmatrix}$ and $B = \begin{bmatrix} 2 & 5 \\ 4 & -1 \\ 1 & 3 \end{bmatrix}$.

We have $AB = \begin{bmatrix} 14 & 25 \\ 17 & 35 \end{bmatrix}$ and therefore $(AB)^T = \begin{bmatrix} 14 & 17 \\ 25 & 35 \end{bmatrix}$.

Now, if we calculate $B^T A^T$, then we have

$$\begin{bmatrix} 4 & 3 \\ 1 & 1 \\ 2 & 7 \end{bmatrix} \begin{bmatrix} 2 & 4 & 1 \\ 5 & -1 & 3 \end{bmatrix} = \begin{bmatrix} 14 & 17 \\ 25 & 35 \end{bmatrix}$$

which illustrates the general result that, for conformable matrices $(AB)^T = B^T A^T$.

REVIEW EXERCISES – SECTION 8.1

1. Calculate the product AB for the following pairs of matrices

(a) $A = \begin{bmatrix} 4 & 3 \\ 2 & 1 \end{bmatrix}$ $\quad B = \begin{bmatrix} 1 & 3 \\ 4 & 6 \end{bmatrix}$

(b) $A = \begin{bmatrix} 3 \\ 4 \end{bmatrix}$ $\quad B = \begin{bmatrix} 2 & 1 \end{bmatrix}$

(c) $A = \begin{bmatrix} 5 & 7 \end{bmatrix}$ $\quad B = \begin{bmatrix} 1 \\ 2 \end{bmatrix}$

2. For the matrices $A = \begin{bmatrix} 1 & 2 \\ 4 & 3 \end{bmatrix}$ and $B = \begin{bmatrix} 1 & 4 \\ 2 & 1 \end{bmatrix}$, show that the transpose of the product is equal to the product of the transposes, that is, $(AB)^T = B^T A^T$.

8.2 DETERMINANTS

> The *determinant* of a matrix is a number which is a function of its elements, and which gives important information about the matrix.

The *determinant* is a unique scalar value that is associated with any square matrix. It provides important information about the nature of the matrix. If the determinant is nonzero, then the matrix is said to be *nonsingular*, which means that the rows and the columns of the matrix are linearly independent. If the determinant is equal to zero, then the matrix is said to be *singular*. In the case of a 2×2 matrix, the determinant is computed as the product of the diagonal elements minus the product of the off-diagonal elements. That is, we have

$$\det(A) = \begin{vmatrix} a_{11} & a_{12} \\ a_{21} & a_{22} \end{vmatrix} = a_{11}a_{22} - a_{12}a_{21}. \tag{8.3}$$

Equation (8.3) illustrates some standard notation for determinants which can be written as either $\det(A)$ or $|A|$ where A is the matrix of interest.

EXAMPLE

Calculate the determinant of the matrix $A = \begin{bmatrix} 4 & 2 \\ 1 & 3 \end{bmatrix}$.

In this case we have $\det(A) = 4 \times 3 - 2 \times 1 = 10$.

For a square matrix A, the property $\det(A) \neq 0$ immediately establishes that its rows and columns are linearly independent. Alternatively, we can say that the matrix A is of *full rank*. That is, the number of independent rows (or columns) is equal to the dimension of the matrix. Conversely, if $\det(A) = 0$, then the matrix A is of *less than full rank*. That is the rows or columns of the matrix are linearly dependent.

EXAMPLE

Calculate the determinant of the matrix $A = \begin{bmatrix} 3 & 6 \\ 1 & 2 \end{bmatrix}$.

We have $\det(A) = 3\times 2 - 6\times 1 = 0$, which demonstrates that the matrix A is singular. If we examine the matrix closely, we see that the second column is equal to the first column multiplied by two. Therefore, the columns are not linearly independent.

The calculation of the determinant becomes more complicated when we consider matrices of higher dimensions. To define the determinant for square matrices of dimension 3×3 and higher, we must first define the *minors* of a square matrix. The *i*, *j*th minor of a matrix A is defined as the determinant of the submatrix obtained by deleting the *i*th row and the *j*th column of A and is written M_{ij}. An associated scalar value is the *i*, *j*th *cofactor* which is equal to $C_{ij} = (-1)^{i+j} M_{ij}$.

EXAMPLE

Consider the matrix $A = \begin{bmatrix} 1 & 4 & 2 \\ 3 & 1 & 4 \\ 5 & 2 & 7 \end{bmatrix}$. The matrix A has a total of nine minors and associated cofactors. Those based on the first row can be calculated as follows.

$$M_{1,1} = \begin{vmatrix} 1 & 4 \\ 2 & 7 \end{vmatrix} = -1 \qquad C_{1,1} = (-1)^2 M_{ij} = -1$$

$$M_{1,2} = \begin{vmatrix} 3 & 4 \\ 5 & 7 \end{vmatrix} = 1 \qquad C_{1,2} = (-1)^3 M_{ij} = -1$$

$$M_{1,3} = \begin{vmatrix} 3 & 1 \\ 5 & 2 \end{vmatrix} = 1 \qquad C_{1,3} = (-1)^4 M_{ij} = 1.$$

We can define matrices of minors and cofactors as shown below for this example.

$$M = \begin{bmatrix} -1 & 1 & 1 \\ 24 & -3 & -18 \\ 14 & -2 & -11 \end{bmatrix} \qquad C = \begin{bmatrix} -1 & -1 & 1 \\ -24 & -3 & -18 \\ 14 & 2 & -11 \end{bmatrix}.$$

The determinant of the matrix can be defined in terms of its minors, or its cofactors, using any row or column. In our example, using the first row, we have

$$\det(A) = \sum_{j=1}^{3} a_{1j} (-1)^{1+j} M_{1j} = \sum_{j=1}^{3} a_{1j} C_{1j}.$$

The process of calculating the determinant in this way is referred to as *expanding* along row one. More generally, we can expand along any of the three rows of the matrix to calculate the determinant as

$$\det(A) = \sum_{j=1}^{3} a_{ij}(-1)^{i+j} M_{ij} = \sum_{j=1}^{3} a_{ij} C_{ij} \qquad \text{for } i = 1,2,3$$

Alternatively, we can expand along any of the three columns of the matrix to obtain

$$\det(A) = \sum_{i=1}^{3} a_{ij}(-1)^{i+j} M_{ij} = \sum_{i=1}^{3} a_{ij} C_{ij} \qquad \text{for } j = 1,2,3.$$

The value of the determinant we calculate does not depend on which row or column we use for the calculation. We cannot prove this statement at this stage, but we can illustrate it by example.

EXAMPLE

The determinant of the matrix $A = \begin{bmatrix} 1 & 4 & 2 \\ 3 & 1 & 4 \\ 5 & 2 & 7 \end{bmatrix}$ can be calculated as

$$\begin{aligned} \det(A) &= 1 \times (-1)^2 \times \begin{vmatrix} 1 & 4 \\ 2 & 7 \end{vmatrix} + 4 \times (-1)^{1+2} \begin{vmatrix} 3 & 4 \\ 5 & 7 \end{vmatrix} + 2 \times (-1)^{1+3} \begin{vmatrix} 3 & 1 \\ 5 & 2 \end{vmatrix} \\ &= -1 - 4 + 2 \\ &= -3. \end{aligned}$$

We have chosen to expand along row 1 to calculate the determinant, but would we have gotten a different answer if we had chosen row 2 or row 3, or indeed column 1, 2, or 3? The answer is no. To illustrate this, consider what would happen if we expanded using column 3. We have

$$\begin{aligned} \det(A) &= 2 \times (-1)^4 \times \begin{vmatrix} 3 & 1 \\ 5 & 2 \end{vmatrix} + 4 \times (-1)^5 \times \begin{vmatrix} 1 & 4 \\ 5 & 2 \end{vmatrix} + 7 \times (-1)^6 \begin{vmatrix} 1 & 4 \\ 3 & 1 \end{vmatrix} \\ &= 2 + 72 - 77 \\ &= -3. \end{aligned}$$

If you are not satisfied, then try expanding along any of the other rows or columns. You will get the same answer.

The property that the choice of row or column is irrelevant for the calculation of the determinant can prove to be a significant advantage when we have matrices in which some rows or columns have several zero elements. If this is the case, then we can often simplify the calculation of the determinant by a careful choice of row or column along which to expand.

EXAMPLE

Calculate the determinant of the following matrix $A = \begin{bmatrix} 1 & 0 & 2 \\ 3 & 4 & 1 \\ 7 & 0 & 5 \end{bmatrix}$.

We note that the second column contains one only nonzero element. Therefore, if we use column 2 for the calculation of the determinant, we need only calculate one minor for the matrix. We have

$$\begin{aligned} \det(A) &= 4 \times (-1)^4 \begin{vmatrix} 1 & 2 \\ 7 & 5 \end{vmatrix} \\ &= 4 \times (5 - 14) \\ &= -36. \end{aligned}$$

We would have obtained the same answer if we had expanded along either of the other two columns or any of the three rows. However, all these choices would have involved calculating three minors rather than the single minor required for this choice.

You will have already noticed that the calculation of the determinant for a 3×3 matrix involves significantly more intermediate calculations than was the case for a 2×2 matrix. If we increase the dimension of the matrix further, then the number of calculations involved increases even more. However, the methods involved in the calculation do not change. Consider a square matrix of dimension n. The general formula for calculation of the determinant by expansion along the ith row can be written as

$$\det(A) = \sum_{j=1}^{n} a_{ij} (-1)^{i+j} M_{ij} = \sum_{j=1}^{n} a_{ij} C_{ij} \qquad \text{for } i = 1, 2, \ldots, n$$

or, for expansion along the jth column, we have

$$\det(A) = \sum_{i=1}^{n} a_{ij} (-1)^{i+j} M_{ij} = \sum_{i=1}^{n} a_{ij} C_{ij} \qquad \text{for } j = 1, 2, \ldots, n.$$

As n increases, the number of intermediate calculations increases rapidly. For example, if $n = 4$, then each of the minors is the determinant of a 3×3 matrix, and we have already seen that this can involve a large number of calculations. We can sometimes cut down on the work involved by a careful choice of row or column for the calculation, but this is not always possible. Fortunately, computers excel at this task and we will normally rely on computers to calculate the determinant for higher-order matrices.

Two useful properties of a determinant are

1. The determinant of the transpose of a matrix is equal to the determinant of the original matrix $\det(A^T) = \det(A)$.
2. The determinant of the product of two square matrices is equal to the product of their determinants $\det(AB) = \det(A)\det(B)$.

Both these properties are stated without proof because the proofs, while not difficult, require many steps.

REVIEW EXERCISES – SECTION 8.2

1. For a general 2×2 matrix A, show that the determinant will be zero if the second column is a multiple of the first column.
2. For the matrix $A = \begin{bmatrix} 1 & 4 & 2 \\ 3 & 1 & 4 \\ 5 & 2 & 7 \end{bmatrix}$, show that the values of the determinant obtained when we expand along the second row, or the first column, are both equal to –3.

8.3 MATRIX INVERSION

> The inverse of a matrix provides a method for the solution of systems of linear equations. It can only be calculated for nonsingular square matrices.

In many situations, it is useful to be able to compute the *matrix inverse*. We define the matrix inverse as follows. Let I be the *identity matrix*, that is a square matrix which has ones on its diagonal and zeros elsewhere. For any square matrix A, the identity matrix has the property that $AI = A$, that is multiplication of the matrix A by the identity matrix simply returns the original matrix. Now, if we can find a matrix B such that $BA = I$, then this defines the inverse of matrix A. The matrix inverse of A is normally written A^{-1}. Note that the matrix inverse is only defined for square matrices and only exists if the rows and columns of A are linearly independent.

Let us begin with the simple case of a 2×2 matrix. We can write the general form of such a matrix as

$$A = \begin{bmatrix} a_{11} & a_{12} \\ a_{21} & a_{22} \end{bmatrix}. \tag{8.4}$$

It is straightforward to show that, if the matrix inverse exists, then it takes the form

$$A^{-1} = \frac{1}{\Delta}\begin{bmatrix} a_{22} & -a_{12} \\ -a_{21} & a_{11} \end{bmatrix}$$

$$\Delta = a_{11}a_{22} - a_{21}a_{22}. \tag{8.5}$$

The proof of inverse form is left as one of the end-of-section exercises for the interested reader. This form also establishes the condition that the matrix must be nonsingular for its inverse to exist, that is, we must have $\Delta \neq 0$, where Δ is the determinant. This condition holds if the matrix has rows and columns which are linearly independent.

EXAMPLE

Let $A = \begin{bmatrix} 4 & 3 \\ 2 & 1 \end{bmatrix}$. The matrix A has determinant $\Delta = 4\times1 - 3\times2 = -2$ and therefore its inverse exists. The inverse can be calculated as

$$A^{-1} = \frac{1}{-2}\begin{bmatrix} 1 & -3 \\ -2 & 4 \end{bmatrix} = \begin{bmatrix} -0.5 & 1.5 \\ 1 & -2 \end{bmatrix}.$$

We can check that this is the inverse matrix by premultiplying A by A^{-1} to show that we get the identity matrix. In this case, we have

$$\begin{bmatrix} -0.5 & 1.5 \\ 1 & -2 \end{bmatrix}\begin{bmatrix} 4 & 3 \\ 2 & 1 \end{bmatrix} = \begin{bmatrix} 1 & 0 \\ 0 & 1 \end{bmatrix}$$

which confirms that our calculations are correct.

The calculation of the matrix inverse for higher-order matrices is less straightforward and will involve a great deal more calculation. Let us consider a square matrix A with dimension n, that is it has n rows and n columns. The inverse of the matrix can be defined in terms of its cofactors as

$$A^{-1} = \frac{1}{\det(A)} \begin{bmatrix} C_{1,1} & C_{2,1} & \cdots & C_{n,1} \\ C_{1,2} & C_{2,2} & & \\ \vdots & & \ddots & \\ C_{1,n} & & & C_{n,n} \end{bmatrix}. \tag{8.6}$$

We can show that the expression for the inverse of a 2×2 matrix given in (8.5) is a special case of this more general expression. The proof of the general result is beyond the scope of this book, but we will illustrate it using some examples.

EXAMPLE

Find the inverse of the matrix $A = \begin{bmatrix} 1 & 4 & 2 \\ 3 & 1 & 6 \\ 1 & 2 & 3 \end{bmatrix}$.

We can calculate the inverse of the matrix A as follows. First, we calculate the matrix M which consists of the minors of A. That is the i, jth element is $\det(A_{i,j})$ where $A_{i,j}$ is the submatrix obtained by deleting row i and column j from A. This gives us

$$M = \begin{bmatrix} -9 & 3 & 5 \\ 8 & 1 & -2 \\ 22 & 0 & -11 \end{bmatrix}.$$

Next, we find the matrix of cofactors where $C_{i,j} = (-1)^{i+j} M_{i,j}$. The matrix of cofactors is therefore given by

$$C = \begin{bmatrix} -9 & -3 & 5 \\ -8 & 1 & 2 \\ 22 & 0 & -11 \end{bmatrix}.$$

We can find the determinant of A by expanding along any row or column of the matrix of cofactors. For example, expanding along the first row gives us

$$\det(A) = 1\times(-9) - 4\times 3 + 2\times 5 = -11.$$

Transposing the matrix of cofactors and dividing by the determinant gives us the inverse of A. We have

$$A^{-1} = \begin{bmatrix} 9/11 & 8/11 & -2 \\ 3/11 & -1/11 & 0 \\ -5/11 & -2/11 & 1 \end{bmatrix}.$$

Although this method is very general, and can, in principle, be applied to any nonsingular square matrix, it becomes very computationally expensive for matrices with higher dimensions. For a 3×3 matrix we needed to calculate 9 minors but, for a 4×4 matrix we would have to calculate 16 minors, each of which is the determinant of a 3×3 matrix and hence, itself requires 9 minors. As we increase the dimension of the matrix, the number of minors we need to calculate is proportional to factorial n. After a while, this starts to become a problem even for computers. Fortunately, there are more efficient numerical algorithms we can use to calculate the matrix inverse which we will discuss in the next section.

REVIEW EXERCISES – SECTION 8.3

1. Show that the general 2×2 matrix $\begin{bmatrix} a_{11} & a_{12} \\ a_{21} & a_{22} \end{bmatrix}$ has inverse $\frac{1}{\Delta}\begin{bmatrix} a_{22} & -a_{12} \\ -a_{21} & a_{11} \end{bmatrix}$ where $\Delta = a_{11}a_{22} - a_{12}a_{21}$.
2. Using the general method given in the text, find the inverse of the matrix A where

$$A = \begin{bmatrix} 2 & 3 & 1 \\ 1 & 2 & 2 \\ 3 & 1 & 1 \end{bmatrix}.$$

8.4 SOLVING SIMULTANEOUS EQUATIONS WITH MATRICES

> Matrix methods are useful in the solution of systems of simultaneous equations. To use these methods, we need to find efficient ways to solve for the matrix inverse.

In this section, we will show how matrix methods can be used to solve systems of linear simultaneous equations. These are systems of equations which can be written in the form $A\mathbf{x} = b$, where $\mathbf{x}$ is a vector of unknown variables, A is a matrix of coefficients and b is a vector of parameters.

EXAMPLE

Consider the system of linear simultaneous equations

$$\begin{aligned} x + 3y + 5z &= 2 \\ 4x + 2y + z &= 1. \\ 2x + y + 3z &= 2 \end{aligned}$$

We can write this system in matrix form as

$$\begin{bmatrix} 1 & 3 & 5 \\ 4 & 2 & 1 \\ 2 & 1 & 3 \end{bmatrix} \begin{bmatrix} x \\ y \\ z \end{bmatrix} = \begin{bmatrix} 2 \\ 1 \\ 2 \end{bmatrix} \tag{8.7}$$

where $\mathbf{x} = \begin{bmatrix} x & y & z \end{bmatrix}^T$ is the vector of unknown variables, A is the matrix of coefficients, and the b is a vector of constants.

Now, if we can find the inverse of the matrix A, then the solution of this system is straightforward. We simply premultiply both sides of the matrix equation (8.7) by A^{-1} to obtain $\mathbf{x} = A^{-1}b$. In this case, we can solve for the inverse of A using the method given in Section 8.3. This gives us a solution of the form

$$\begin{bmatrix} x \\ y \\ z \end{bmatrix} = \begin{bmatrix} -0.2 & 0.16 & 0.28 \\ 0.4 & 0.28 & -0.76 \\ 0 & -0.2 & 0.4 \end{bmatrix} \begin{bmatrix} 2 \\ 1 \\ 2 \end{bmatrix} = \begin{bmatrix} 0.32 \\ -0.44 \\ 0.6 \end{bmatrix}.$$

This method is fine when we have a small number of equations in the system but, as the number of equations increases, the number of calculations involved in calculating the matrix inverse increases much faster. For systems with many equations, this method becomes computationally expensive, and it is useful to look for alternative methods to solve such systems.

The first method we will consider is *Cramer's rule*. This method is useful when we are only interested in solving for a subset of the unknown variables. If this is the case, then we can economize on the number of calculations by focusing on these variables only. Consider the system $A\mathbf{x}=b$ where $\mathbf{x}$ is an $n\times 1$ vector, A is an $n\times n$ matrix, and b is an $n\times 1$ vector. Now suppose we are only interested in solving for x_1 which is the first element of the vector $\mathbf{x}$. Cramer's rule states that the solution for this element is given by the expression

$$x_1 = \frac{\det(A_1)}{\det(A)}$$

where A_1 is the matrix obtained by substituting the vector b for the first column of A.

EXAMPLE

For the system (8.7), we have $\det(A)=-25$. Substituting b for the first column of A, we have

$$A_1 = \begin{bmatrix} 2 & 3 & 5 \\ 1 & 2 & 1 \\ 2 & 1 & 3 \end{bmatrix}.$$

It is straightforward to calculate the determinant of A_1 as $\det(A_1)=-8$ and, by Cramer's rule, it follows that $x_1=\det(A_1)/\det(A)=(-8)/(-25)=0.32$. This is the same solution we obtained by premultiplying the vector b by the matrix A^{-1}.

Note that Cramer's rule can be generalized to solve for any of the elements of the vector of variables $\mathbf{x}$. We have $x_i=\det(A_i)/\det(A)$ for $i=1,\ldots,n$, where A_i is the matrix obtained by replacing the ith column of A with the vector b.

Cramer's rule is often useful when solving systems of equations derived from economic theory. In such cases, we often wish to solve for the equilibrium of a system which is defined in terms of symbols rather than specific numerical values. This is generally very easy to do using Cramer's rule as we will illustrate using the following example.

EXAMPLE

Consider the open-economy income-expenditure model of national output defined by the following equations

$$\begin{aligned} Y &= C + I + G + X - M \\ C &= b + cY \\ M &= d + eY \end{aligned}$$

where Y is national output, C is consumption, I is investment, G is government spending, X is exports, and M is imports. This system can be written in matrix form as

$$\begin{bmatrix} 1 & -1 & 1 \\ -c & 1 & 0 \\ -e & 0 & 1 \end{bmatrix} \begin{bmatrix} Y \\ C \\ M \end{bmatrix} = \begin{bmatrix} I+G+X \\ b \\ d \end{bmatrix}.$$

Let us define A to be the matrix on the left-hand side of this expression. We can calculate the determinant of A by expanding along row 2 to obtain $\det(A) = -c + 1 + e$. This allows us to solve for Y in terms of the exogenous variables and the parameters of the model using Cramer's rule. First, we replace the first column of A with the right-hand side vector to obtain

$$A_1 = \begin{bmatrix} I+G+X & -1 & 1 \\ b & 1 & 0 \\ d & 0 & 1 \end{bmatrix}.$$

Expanding along row 2 gives us $\det(A_1) = b + I + G + X - d$. Therefore, by Cramer's rule, we can solve for national output as $Y = \det(A_1)/\det(A)$ which gives us

$$Y = \frac{b + I + G + X - d}{1 - c + e}.$$

This is a familiar equation in macroeconomic theory which shows that the level of national output is the product of the total level of autonomous expenditure $(b+I+G+X-d)$ and the multiplier $1/(1-c+e)$.

Cramer's rule avoids the problem of inverting the A matrix in the system $\mathbf{Ax}=b$ by concentrating on a subset of variables for which we need to find the solution. If, however, we need to solve for all the unknown variables, then it is not an efficient way to solve the model. A better alternative in these circumstances is to look for more efficient ways to invert the A matrix to obtain a full solution of the model. One such method is the use of the *LU decomposition*. This provides a particularly useful method which is widely used in many computer applications. It works as follows:

1. For the matrix A, find the matrices L and U such that $LU=A$ and where L is lower triangular, and U is upper triangular.[1]
2. Solve for the matrix Y such that $LY=I$.
3. Solve for the matrix X such that $UX=Y$. The matrix $X=A^{-1}$ is the inverse of the original matrix A.

Stage 1 is achieved by a sequence of row operations on the matrix of interest. Once the LU decomposition has been found, stages 2 and 3 are straightforward. Stage 2 is implemented using the method of forward substitution, which is possible in this case because the matrix L is lower triangular. Similarly, stage 3 is implemented using the method of backward substitution, which is possible because the matrix U is upper triangular. Note that the process of finding the LU decomposition of a matrix has much in common with the method of Gaussian elimination which we discussed in Chapter 3. Although it is possible to use this algorithm to find the inverse of matrices by hand, it can involve a lot of tedious calculations. However, it can be implemented as a very efficient computer algorithm for inverting higher dimension matrices. Code for this method is shown in Figure 8.2. Note that this requires the input of the dimension n and the matrix A for the program to run.

---

[1] The *principal diagonal* of a square matrix consists of the elements which run from the top left to the bottom right. A lower triangular matrix is one in which all elements below the principal diagonal are equal to zero. An upper triangular matrix is one in which all elements above the principal diagonal are equal to zero.

```
n = 4

a = np.zeros((n,n))
a = ([4,3,1,2],[7,4,9,1],[5,2,3,7],[4,6,8,1])
umat = a
lmat = np.zeros((n,n))
ymat = np.zeros((n,n))
c = np.zeros((n,n))

for idx in range(0,n):
    c[idx][idx]=1
    lmat[idx][idx]=1

for jdx in range(0,n):
    for idx in range(jdx+1,n):
        lmat[idx][jdx]=umat[idx][jdx]/umat[jdx][jdx]
        for kdx in range(0,n):
            umat[idx][kdx]=umat[idx][kdx]-lmat[idx][jdx]*umat[jdx][kdx]

y = np.zeros((n,n))

for jdx in range(0,n):
    y[0][jdx]=c[0][jdx]/lmat[0][0]
    for idx in range(1,n):
        sum1 = c[idx,jdx]
        for kdx in range(0,idx):
            sum1=sum1-lmat[idx][kdx]*y[kdx][jdx]
        y[idx][jdx]=sum1/lmat[idx][idx]

ainv = np.zeros((n,n))

for jdx in range(n-1,-1,-1):
    ainv[n-1][jdx]=y[n-1][jdx]/umat[n-1][n-1]
    for idx in range(n-1,-1,-1):
        sum1 = y[idx][jdx]
        for kdx in range(n-1,idx,-1):
            sum1=sum1-umat[idx][kdx]*ainv[kdx,jdx]
        ainv[idx,jdx]=sum1/umat[idx][idx]

print("The matrix inverse is ")
print()
print(ainv)
```

FIGURE 8.2 Python Code for Matrix Inversion by LU factorization.

EXAMPLE

Using the code in Figure 8.2, we will solve for the inverse of the 4×4 matrix

$$A=\begin{bmatrix}4 & 3 & 1 & 2\\ 7 & 4 & 9 & 1\\ 5 & 2 & 3 & 7\\ 4 & 6 & 8 & 1\end{bmatrix}.$$

First, we note that the LU factorization of this matrix gives us the following lower triangular L matrix, and upper triangular U matrix.

$$L=\begin{bmatrix}1 & 0 & 0 & 0\\ 1.75 & 1 & 0 & 0\\ 1.25 & 1.4 & 1 & 0\\ 1 & -2.4 & -2.9048 & 1\end{bmatrix}\quad U=\begin{bmatrix}4 & 3 & 1 & 2\\ 0 & -1.25 & 7.25 & -2.5\\ 0 & 0 & -8.4 & 8\\ 0 & 0 & 0 & 16.2381\end{bmatrix}.$$

We can then use these to solve the matrix systems $LY=I$ and $UX=Y$ by forward and backward substitution, to give us the inverse matrix $A^{-1}=X$.

$$A^{-1}=\begin{bmatrix}0.2141 & 0.1804 & -0.0572 & -0.2082\\ 0.1994 & -0.1950 & -0.0601 & 0.2170\\ -0.2434 & 0.0689 & 0.0513 & 0.0587\\ -0.1056 & -0.1026 & 0.1789 & 0.0616\end{bmatrix}.$$

Once we have solved for the matrix inverse, it becomes straightforward to solve for the vector $\mathbf{x}$ in the expression $A\mathbf{x}=b$ as $\mathbf{x}=A^{-1}b$.

EXAMPLE

1. For $b=\begin{bmatrix}1 & 1 & 1 & 1\end{bmatrix}^T$ we have $\mathbf{x}=\begin{bmatrix}0.1290 & 0.1613 & -0.0645 & 0.0323\end{bmatrix}^T$
2. For $b=\begin{bmatrix}1 & 0 & 0 & 0\end{bmatrix}^T$ we have $\mathbf{x}=\begin{bmatrix}0.2141 & 0.1994 & -0.2434 & -0.1056\end{bmatrix}^T$
3. For $b=\begin{bmatrix}0 & 1 & 0 & 0\end{bmatrix}^T$ we have $\mathbf{x}=\begin{bmatrix}0.1804 & -0.1950 & 0.0689 & -0.1026\end{bmatrix}^T$

REVIEW EXERCISES – SECTION 8.4

1. Using the computer code provided, find the inverse of the matrix

$$A = \begin{bmatrix} 2 & 1 & 3 & 0 \\ 4 & 1 & 0 & 5 \\ 0 & 3 & 5 & 7 \\ 1 & 3 & 6 & 9 \end{bmatrix}$$

2. Consider the following model of demand and supply for two goods in related markets.

$$\begin{aligned} q_1^s &= 25 + 2p_1 \\ q_2^s &= 50 + p_2 \\ q_1^d &= 100 - 0.5p_1 + 0.25p_2 \\ q_2^s &= 150 + 0.5p_1 - 0.75p_2. \end{aligned}$$

   (a) Write the model in matrix form.

   (b) Use the matrix form of the model to solve for the equilibrium price and quantity in each market.

8.5 EIGENVALUES AND EIGENVECTORS

> *Eigenvalues* and *eigenvectors* give us information about the properties of a square matrix. They are particularly useful when solving systems of difference or differential equations.

Eigenvalues are scalar values associated with a square matrix, and *eigenvectors* are vectors which are associated with these values. Eigenvalues are also referred to as the *roots* or *characteristic values* of the matrix. We can define an eigenvalue of the matrix A as any value λ such that $Ax = \lambda x$ for a nonzero vector x. The vector x is the eigenvector associated with λ.

To solve for the eigenvalues of the matrix A, we note that, if $Ax = \lambda x$, then we can write

$$(A - \lambda I)x = 0 \qquad (8.8)$$

where 0 is a vector of zeros. If x is a vector that contains at least one nonzero element, then (8.8) can only be true if the matrix $(A-\lambda I)$ is singular. That is, we require $\det(A-\lambda I)=0$. This condition is the definition of an eigenvalue. For an $n\times n$ matrix A, the condition $\det(A-\lambda I)=0$ defines a polynomial function of λ of order n. This equation is referred to as the *characteristic equation* of the matrix. It follows that there are n possible solutions to the characteristic equation, but these may not be unique.

EXAMPLE

Consider the matrix $A=\begin{bmatrix}0.5 & -0.5\\ 1.5 & 2.5\end{bmatrix}$.

The eigenvalues are defined by the condition $\begin{vmatrix}0.5-\lambda & -0.5\\ 1.5 & 2.5-\lambda\end{vmatrix}=0$ which gives us the characteristic equation $\lambda^2-3\lambda+2=0$. This equation factorizes easily to give us $(\lambda-2)(\lambda-1)=0$ and the eigenvalues are, therefore, $\lambda_1=1$ and $\lambda_2=2$.

To solve for the eigenvector associated with $\lambda_1=1$ we look for a vector $x=\begin{bmatrix}x_1 & x_2\end{bmatrix}^T$ such that

$$\begin{bmatrix}0.5 & -0.5\\ 1.5 & 2.5\end{bmatrix}\begin{bmatrix}x_1\\ x_2\end{bmatrix}=\begin{bmatrix}x_1\\ x_2\end{bmatrix}.$$

Using either row of this expression gives us a relationship of the form $x_1=-x_2$. This defines the eigenvector for $\lambda_1=1$. Note that the eigenvector is only determined up to a multiplicative constant. For example, we could set $x_1=1$, which gives us an eigenvector of the form $x=\begin{bmatrix}1 & -1\end{bmatrix}^T$. Another convention is to choose a scaling such that the modulus of the elements is equal to one, that is, $x_1^2+x_2^2=1$ which, in this case, gives us the eigenvector $x=\begin{bmatrix}0.7071 & -0.7071\end{bmatrix}^T$.

To find the eigenvector associated with $\lambda_2=2$, we look for values of x_1 and x_2 which satisfy the expression

$$\begin{bmatrix}0.5 & -0.5\\ 1.5 & 2.5\end{bmatrix}\begin{bmatrix}x_1\\ x_2\end{bmatrix}=\begin{bmatrix}2x_1\\ 2x_2\end{bmatrix}.$$

Using either row we obtain a relationship of the form $x_2=-3x_1$. We can again normalize this in different ways. For example, we could set $x_2=1$ to get an eigenvector of the form $x=\begin{bmatrix}-1/3 & 1\end{bmatrix}^T$. Alternatively, we can set the modulus equal to one which gives us $x=\begin{bmatrix}-0.3162 & 0.9487\end{bmatrix}^T$.

In the case of a 2×2 matrix, there is a useful relationship between the eigenvalues and the trace and determinant. Consider the matrix general 2×2 matrix as defined in equation (8.4). We solve for the eigenvalues by solving the characteristic equation $\lambda^2-(a_{11}+a_{22})\lambda+(a_{11}a_{22}-a_{12}a_{21})=0$ and therefore the solutions take the form

$$\lambda_{1,2}=\frac{(a_{11}+a_{22})\pm\sqrt{(a_{11}+a_{22})^2-4(a_{11}a_{22}-a_{12}a_{21})}}{2}. \tag{8.9}$$

Since the trace of the matrix is defined as the sum of its diagonal elements $tr(A)=a_{11}+a_{22}$, and its determinant is defined as $\det(A)=a_{11}a_{22}-a_{12}a_{21}$, it follows that we can write the eigenvalues as

$$\lambda_{1,2}=\frac{tr(A)\pm\sqrt{tr(A)^2-4\det(A)}}{2}.$$

Two useful properties are immediately obvious from this expression.

1. The trace of the matrix is equal to the sum of its eigenvalues, $\lambda_1+\lambda_2=tr(A)$.
2. The determinant of the matrix is equal to the product of its eigenvalues, $\lambda_1\lambda_2=\det(A)$.

The proofs of these statements are left as an exercise for the interested reader. Note that these results hold for both real and complex eigenvalues. From these properties, we can derive several other useful results, which are listed below:

1. If $4\det(A)<tr(A)^2$, then the eigenvalues are real and distinct.
2. If $4\det(A)>tr(A)^2$, then the eigenvalues are complex conjugates.
3. If $4\det(A)=tr(A)^2$, then the eigenvalues are real and repeated.
4. If the determinant is negative, then the eigenvalues are real and have opposite sign.
5. If the trace is negative and the determinant is positive, then the eigenvalues are either both real and negative or complex conjugates with negative real part.
6. If the trace is positive and the determinant is positive, then the eigenvalues are either both real and positive or complex conjugates with positive real part.

All these properties are straightforward to prove, and the proofs are again left to the reader. The reason we state these properties here is that it is often more important to know the nature of the eigenvalues rather than their specific numerical values. These conditions give us a quick and easy way to check if the eigenvalues are real or complex and if they are positive, negative or of opposite sign. This is often enough to identify the nature of solutions to difference or differential equation models without needing to solve the associated eigenvalue problems explicitly.

EXAMPLE

Consider the matrix $A = \begin{bmatrix} -1 & 4 \\ 2 & -2 \end{bmatrix}$.

We have $tr(A) = -3$ and $\det(A) = -6$. It follows that the eigenvalues are real and have opposite sign. We can confirm this by solving for them explicitly which gives us values $\lambda_1 = 1.3723$ and $\lambda_2 = -4.3723$.

EXAMPLE

Consider the matrix $A = \begin{bmatrix} 2 & -1 \\ 3 & 2 \end{bmatrix}$.

We have $tr(A) = 4$ and $\det(A) = 7$. Since $tr(A)^2 < 4\det(A)$, it follows that the eigenvalues complex conjugates. We can confirm this by solving for them explicitly which gives us values $\lambda_1 = 2 + 1.7321i$ and $\lambda_2 = 2 - 1.7321i$.

REVIEW EXERCISES – SECTION 8.5

1. Find the eigenvalues and the eigenvectors of the matrix $A = \begin{bmatrix} 3 & 1 \\ 0 & 2 \end{bmatrix}$.
2. Show that, for a 2×2 matrix, if the trace is negative and the determinant is positive, then the eigenvalues are either both real and negative or complex conjugates with negative real parts.

CHAPTER 9

FIRST-ORDER DIFFERENTIAL EQUATIONS

In this chapter, we will examine solution procedures for *first-order differential equations*. Equations of this type occur frequently in economics when we consider dynamic adjustment, that is, adjustment through time in response to disequilibrium. First-order differential equations are of the form $dy/dx = f(x,y)$ which contain the first derivative only of the variable of interest y. We will also limit our attention to situations in which there is only one independent variable x. Such an equation is referred to as an *ordinary differential equation* (ODE) to distinguish it from cases that contain more than one independent variable. We refer to the more general case as a *partial differential equation* (PDE). Such equations are much more difficult to solve and will not be considered here. The solution of an ODE is a function of the form $y(x)$ which determines the value of the variable of interest given the value of the independent variable x. There is no unique way to solve ODEs, and, in some cases, they may not be solvable at all. There are, however, several different solution methods available to us which we will cover in this chapter. Some equations may be solvable by several different methods and the choice between them will depend on which is the easiest to implement.

9.1 SEPARABLE DIFFERENTIAL EQUATIONS

As the name suggests, *separable* differential equations apply when we can separate the function $f(x, y)$ by expressing it as a multiple of a function of x and a function of y. Equations of this kind can be solved by the process of integration.

Suppose we have a differential equation which takes the form

$$\frac{dy}{dx} = g(x)h(y). \tag{9.1}$$

Equations of this type are known as *separable* differential equations because the expression on the right-hand side is the product of two separate expressions, one of which contains only the x variable and the other contains only the y variable. This gives us a significant advantage because it allows us to express the equation in a way which will allow us to solve it by integration.

To solve an equation of the form (9.1), we look for a function of the form $y(x)$, that is, a function that determines the value of y for a given value of x. The separability property will make this easier because, providing $h(y)$ is not equal to zero, it allows us to write (9.1) in the form $\left(1/h(y)\right)dy = g(x)dx$. We can now solve our equation by integrating both sides of the transformed equation. That is, our solution is defined by

$$\int \frac{1}{h(y)} dy = \int g(x) dx .$$

EXAMPLE

Consider the first-order differential equation $\frac{dy}{dx} = \frac{4}{x}$. How can we solve this equation to obtain a function of the form $y(x)$?

Dividing both sides by 4 and multiplying by dx, gives us

$$\frac{1}{4} dy = \frac{1}{x} dx .$$

The next stage is to integrate this equation. This gives us

$$\int \frac{1}{4} dy = \int \frac{1}{x} dx \Rightarrow \frac{1}{4} y = \ln(x) + A$$

where A is a constant of integration. Multiplying both sides by 4, gives us the final form of our solution $y = 4\ \ln(x) + C$, where $C = 4A$ is a multiple of the original constant of integration. This is referred to as a *general solution* since it is true for any constant of integration C. Note that we can always check our solution by differentiating it to confirm that we recover the original differential equation. In this case, we have $d\left(4\ln(x) + C\right)/dx = 4/x$ which confirms that we have the correct solution.

A *particular solution*, or *particular integral*, of a differential equation, is a solution in which the constant of integration is assigned a specific value. The usual way in which this is done is through an *initial condition* of some form. In the case of our example, we have a general solution of the form $y(x)=4\ln(x)+C$. If we also know that, $y(1) = 0$, then we can solve for C as $C=-4\ln(1)=0$. We can therefore write the particular solution which is consistent with this initial condition as $y(x) = 4\ln(x)$.

EXAMPLE

Solve the differential equation $dy/dx = xy^2$ where $y = 1$ when $x = 0$.

This is a separable equation, and we can therefore solve it by integration as

$$\int \frac{1}{y^2}dy = \int x\,dx\,,$$

which gives a general solution

$$-\frac{1}{y}=\frac{x^2}{2}+C\,.$$

We can eliminate the constant of integration by using the initial condition, which gives us C = –1 and this, in turn, gives us the particular solution $1/y = 1 - x^2/2$, or

$$y(x)=\frac{2}{2-x^2}\,.$$

You can again check that this solution is correct by differentiating with respect to x, which recovers the original differential equation.

REVIEW EXERCISES – SECTION 9.1

1. Use the method of separation of variables to solve the following differential equations and use the initial condition to eliminate the constant of integration.

a) $\dfrac{dy}{dx}=\dfrac{xy^2}{(1+x)}$ $y(0)=1$

b) $\dfrac{dy}{dx}=e^{-y}(3x-1)$ $y(0)=0$

2. A firm purchases a machine for \$200, and its resale value subsequently declines according to the equation

$$\frac{dp}{dt} = -0.1p + 10$$

where p is the price it will sell at in the resale market. Solve for the resale price as a function of time.

3. Find the particular solution of the differential equation $\frac{dy}{dx} = \exp(-y)\left(2x - \frac{3}{2}\right)$ with initial condition $y(0) = 1$. Show that this solution is valid for all $x \geq 0$.

9.2 FIRST-ORDER LINEAR DIFFERENTIAL EQUATIONS WITH CONSTANT COEFFICIENTS

The class of equations considered in this section is particularly important in economic applications. Here, we show how to solve equations of this type quickly and easily without the need for integration.

First-order linear differential equations with constant coefficients take the general form given in equation (9.2)

$$\frac{dy}{dx} + ay = b\,, \tag{9.2}$$

where a and b are parameters. Equations of this type can be solved by the separation of variables, as shown in the previous section. There is, however, an easier solution method and, because equations of this type are so frequently encountered, we will explain this method in this section.

Let us begin with a modified version of (9.2) in which the parameter b is equal to zero. This gives us an equation of the form $dy/dx = -ay$, which is the general form of a *homogeneous* first-order linear differential equation with a constant coefficient a. We can find the general solution of this equation very easily by separation of variables. This gives us an equation of the form $y_g(x) = C\exp(-ax)$, where C is an arbitrary constant. This provides the form of the solution for all equations of this type, and since they occur so frequently, we often make use of this form directly rather than going through the process of separation of variables. Once we have the general solution, we can then find the particular solution by using an initial condition to solve for the constant of integration in exactly the same way as we saw in the previous section.

EXAMPLE

Find the particular solution of the differential equation $dy / dx = -0.1y$ with the initial condition $y(0) = 2$.

Since this is a first-order linear differential equation with a constant coefficient, we can use the general formula for the solution and write it as

$$y(x) = C\exp(-0.1x).$$

From the initial condition, we have $C = 2$. The particular solution corresponding to this initial condition, therefore, takes the form

$$y(x) = 2\exp(-0.1x)$$

Now let us return our attention to the more general case given by equation (9.2), in which the parameter b not equal to zero. Equations of this type are referred to as nonhomogeneous first-order linear differential equations with constant coefficients. The term nonhomogeneous indicates the presence of a nonzero constant term in the equation. We will show that the general solution of equation (9.2) is equal to sum of the general solution to the associated homogeneous problem, which we call the *complementary function*, and the particular integral given by the solution of the equation corresponding to the case $dy / dx = 0$, which we call the *particular integral*. This means that the general solution for our equation will take the form

$$y_g(x) = C\exp(ax) - \frac{b}{a}. \tag{9.3}$$

Proof: Rather than solving the equation from first principles, we can simply show that differentiating equation (9.3) with respect to x recovers the original differential equation. We have:

$$\frac{dy}{dx} = aC\exp(ax) = a\left(y(x) + \frac{b}{a}\right) = ay(x) + b.$$

This confirms that (9.3) is the general solution for the general differential equation defined in (9.2).[1]

EXAMPLE

Find the general solution of the differential equation $\frac{dy}{dx} = 3y - 2.$

The complementary function is the solution of the associated homogeneous differential equation and is given by $y_c(x) = C\exp(3x)$. The particular integral associated with $dy/dx = 0$ is $y_p = 2/3$. Therefore, the general solution to the equation given is:

$$y_g(x) = C\exp(3x) + \frac{2}{3}.$$

We are not asked to solve for a particular solution here, but the procedure for doing so would be the same as for our earlier examples. That is, we would use an initial condition of some form to solve for the constant of integration.

EXAMPLE

Find the particular solution of the differential equation $\frac{dy}{dx} = -3y + 6$ with the initial condition $y(0) = 12$.

The general solution of this equation is equal to the sum of the complementary function and the particular integral. This gives us

$$y_g(x) = C\exp(-3x) + 2.$$

To solve for the constant of integration, we use the initial condition. This gives us $12 = C\exp(0) + 2$, which, in turn, gives us $C = 10$. Therefore, the particular solution for this equation for the given initial condition is

$$y(x) = 10\exp(-3x) + 2.$$

Next, to illustrate the use of differential equations in economic analysis, let us consider an example in which dynamic market adjustment naturally gives rise to a relationship that is expressed in the form of a differential equation.

---

[1] This is an example of a more general result known as the *principle of superposition* which we will discuss in more detail later.

We will also show how the relationship we derive can be solved to yield an equation in which the price of a good adjusts through time in response to market disequilibrium.

EXAMPLE

Consider a market for a good in which demand is given by the following function of price $q^d = 200 - 2p$ and there is a fixed supply $q^s = 100$. If price adjusts to the gap between demand and supply according to the equation $dp/dt = 0.5(q^d - q^s)$ and $p(0) = 75$, where t is a time index, solve for price as function of time.

Substituting the demand and supply relationships into a price adjustment equation gives us a differential equation of the form

$$\frac{dp}{dt} = 0.5(200 - 2p - 100) = -p + 50 .$$

This is a nonhomogeneous first-order differential equation that can be solved using the general method we have set out. The general solution is given by the sum of the complementary function and the particular integral. This gives us

$$p_g(t) = C\exp(-t) + 50 .$$

Using the initial condition, we have $75 = C\exp(0) + 50$ which solves to give us $C = 25$. Therefore, the particular solution can be written

$$p(t) = 25\exp(-t) + 50.$$

Note that the negative coefficient on t in this equation means that the first term will tend to zero as t becomes large. Therefore, as $t \to \infty$, the price converges on its equilibrium value of 50.

When working with differential equations in the context of economics and business models, we are often concerned with the issue of stability. Most often, differential equations in this context are concerned with modeling adjustment over time, and we are interested in whether the variable of interest converges on a long-run equilibrium. It is relatively easy to check for stability when dealing with first-order equations. For equations of the form (9.2), we can show that if $a > 0$, then the particular solution of the differential equation will tend toward the equilibrium given by the particular integral as x becomes large for any value of the initial condition. In contrast, if $a<0$, then the solution diverges from the equilibrium for any initial condition $y(0)\neq y_p$. Conditions for stability are harder to derive for higher-order differential equations, and we will consider this issue in Chapter 10.

REVIEW EXERCISES – SECTION 9.2

1. Show that the general solution of the differential equation $\frac{dy}{dx}-y=-4$ is the same when we solve it using the method of separation of variables and when we solve it as the sum of the complementary function and the particular integral.

2. Find the general solution of each of the following differential equations
   (a) $\frac{dy}{dx}+0.2y=3$
   (b) $\frac{dy}{dx}=0.1y-100$
   (c) $4\frac{dy}{dx}+2y=6$

3. Find the particular solution for each of the following differential equations using the initial conditions given
   (a) $\frac{dy}{dx}-2y=4$ $\quad y(0)=1$
   (b) $\frac{dy}{dx}+3y=3$ $\quad y(-1)=2$
   (c) $\frac{dy}{dx}+0.1y=2$ $\quad y(1)=5$

9.3 SOLUTIONS USING AN INTEGRATING FACTOR

In this section, we introduce the use of an integrating factor to solve first-order differential equations in which the coefficients are not constant. This method generalizes the problem considered in the previous section by allowing the coefficients in the equation to depend on the value of the independent variable x.

The general form of a first-order linear differential equation can be written.

$$\frac{dy}{dx}+p(x)y=q(x). \tag{9.4}$$

This generalizes the differential equations we considered in the previous section by allowing the coefficients p and q to be functions of x. Note that this is still a linear differential equation because dy/dx is a linear function of y. However, the coefficient on y is given by $p(x)$ and is, therefore, not constant. Moreover, both the functions $p(x)$ and $q(x)$ may be nonlinear functions of x. We will, however, assume that both these functions are continuous and integrable.

We will begin with the homogeneous case, in which we assume that $q(x)=0$. For this case, we can use separation of variables as a solution method. However, the method of integrating factors offers an alternative solution technique that we can extend to the case of nonhomogeneous equations in which $q(x)\neq 0$. To illustrate this new technique, let us begin with an example

EXAMPLE

Suppose we have a differential equation of the form $\frac{dy}{dx}+xy=0$.

To solve this equation, we begin by multiplying through by a function of x given by $v(x)=\exp\left(x^2/2\right)$. This transforms the equation to give

$$\exp\left(\frac{x^2}{2}\right)\frac{dy}{dx}+x\exp\left(\frac{x^2}{2}\right)y=0.$$

$v(x)$ is referred to as the *integrating factor*. At first glance, it might appear that multiplying through by the integrating factor has just made the equation more complicated. We can, in fact, show that this allows us to simplify the equation considerably. If we look carefully at the transformed equation, we see that we can write it in the form

$$\frac{d\left(y\exp\left(x^2/2\right)\right)}{dx}=0.$$

Since this is written in the form of a derivative with no other terms present, integration is now trivial, and we can write down a general solution of the form

$y\exp\left(x^2/2\right)=C$. Multiplying this expression by $\exp\left(-x^2/2\right)$ now allows us to write the explicit form of the general solution as

$$y(x)=C\exp\left(-\frac{x^2}{2}\right).$$

For our example, we have a very straightforward solution method *providing* we know what integrating factor to use to simplify the original differential equation. However, it is not obvious where the function $v(x)$ has come from. We simply introduced it and showed that it worked to solve this particular problem. If we are to use the method more generally, however, we will need to have a more systematic approach in which we identify a method by which we can determine the function $v(x)$.

We can generalize the integrating factor method for homogeneous linear differential equations as follows. Suppose we have $dy/dx+p(x)y=0$ where $p(x)$ is a continuous and integrable function of x. We can now show that the function $v(x)=\exp\left(\int p(x)dx\right)$ will act as a suitable integrating factor in all problems of this type. We can show that

$$v(x)\frac{dy}{dx}+v(x)p(x)y=\frac{d(v(x)y)}{dx}=0$$

when $v(x)$ is defined in this way.

Proof: Let $v(x)=\exp\left(\int p(x)dx\right)$. By the product rule of differentiation we have

$$\frac{d(v(x)y)}{dx}=v(x)\frac{dy}{dx}+\frac{dv(x)}{dx}y. \tag{9.5}$$

To find $dv(x)/dx$, we will make use of the chain rule. Let $u=\int p(x)dx$. Using this, we have $v(u)=\exp(u)$, and we can write

$$\frac{dv(x)}{dx}=\frac{dv(u)}{du}\frac{du}{dx}=\exp(u)p(x)=\exp\left(\int p(x)dx\right)p(x)=v(x)p(x).$$

Substituting this expression into gives us

$$\frac{d(v(x)y)}{dx}=v(x)\frac{dy}{dx}+v(x)p(x)y$$

which establishes that choosing $v(x)$ as the integrating factor will allow us to simplify the differential equation for any continuous and integrable function $p(x)$.

EXAMPLE

Find the particular solution for the differential equation $\frac{dy}{dx}+3x^2y=0$ with initial condition $y(0) = 1$.

We can solve for the integrating factor as $v(x)=\exp\left(\int 3x^2dx\right)=\exp\left(x^3\right)$. This allows us to write the differential equation in the form

$$\frac{d\left(y\exp\left(x^3\right)\right)}{dx}=0$$

which integrates to give $y\exp\left(x^3\right)=C$. The initial condition is now used to solve for C, we have $1\exp(0)=C$ or $C = 1$. Thus, the particular solution for this problem is given by $y(x)=1/\exp\left(x^3\right)$.

The integrating factor method can also be used to solve nonhomogeneous linear differential equations, even though these functions will typically be non-separable. In doing so, we use the same integrating factor as we would for the associated homogeneous problem. That is, for an equation of the form (9.4), we use $v(x)=\exp\left(\int p(x)dx\right)$ as the integrating factor. To illustrate this, let us consider an example.

EXAMPLE

Find the general solution of the nonhomogeneous differential equation $\frac{dy}{dx}+\left(\frac{1}{x}\right)y=3x$ using the integrating factor method.

In this case, we use $v(x)=\exp\left(\int\frac{1}{x}dx\right)$ as the integrating factor. This gives us $v(x)=x$. Multiplying through by the integrating factor transforms the differential equation to $x\,dy/dx+y=3x^2$ or $d(yx)/dx=3x^2$. Integrating this expression yields $yx=x^3+C$. Dividing through by x now yields the following general solution for our differential equation.

$$y_g(x)=x^2+\frac{C}{x}.$$

EXAMPLE

Find the particular solution of the nonhomogeneous differential equation $\frac{dy}{dx} + 2y = \exp(x)$ with initial condition $y(0) = 2$.

The integrating factor for this problem is $v(x) = \exp\left(\int 2\,dx\right) = \exp(2x)$.

Multiplying through transforms the differential equation to $\exp(2x)dy/dx + 2\exp(2x)y = \exp(3x)$ or $d(y\exp(2x))/dx = \exp(3x)$. Therefore, we can write the general solution as

$$y_g(x)\exp(2x) = \frac{1}{3}\exp(3x) + C.$$

From the initial condition, we have $2\exp(0) = \exp(0)/3 + C$, which gives us $C = 5/3$. The particular solution takes the form

$$y(x) = \frac{1}{3}\exp(x) + \frac{5}{3}\exp(-2x).$$

REVIEW EXERCISES – SECTION 9.3

1. Find the general solution for the following homogeneous differential equations using the integrating factor method
   - **(a)** $\frac{dy}{dx} + 0.5y = 0$
   - **(b)** $\frac{dy}{dx} + \left(\frac{4}{x}\right)y = 0$
   - **(c)** $x^2\frac{dy}{dx} + 5y = 0$
2. Using the integrating factor method, solve the following differential equation $\frac{dy}{dx} + 3\frac{y}{x} = x^2$ with initial condition $y(1) = 0$.
3. Using the integrating factor method, show that the general solution of the equation

$$\frac{dy}{dx} + ay = b$$

takes the form $y_g(x) = C\exp(-ax) + \frac{b}{a}$.

9.4 THE METHOD OF UNDETERMINED COEFFICIENTS

The method of undetermined coefficients is particularly important for the solution of nonhomogeneous differential equations. Although it useful in the context of first-order equations, it will become even more useful when we consider higher-order problems.

In Section 9.2, we showed that the general solution to a nonhomogeneous differential equation could be written as the sum of the general solution to the associated homogeneous equation (the complementary function) and any particular solution of the nonhomogeneous equation (the particular integral). By using this property, we were able to solve nonhomogeneous problems with constant coefficients, like problems of the form given in equation (9.2). In this section, we generalize this insight and show that we can solve problems in which the right-hand side of our differential equation is a function of the independent variable x. Initially, however, we will maintain the assumption of a fixed coefficient on the y variable. Although it is possible to consider more general cases, this is considerably more difficult.

The general form for the equations we consider in this section is

$$\frac{dy}{dx} + ay = f(x), \tag{9.6}$$

and we look for a solution of the form

$$y(x) = y_g(x) + y_p(x),$$

where $y_g(x)$ is the general solution of the associated homogeneous equation and $y_p(x)$ is any particular solution. Note that this method will only work if $f(x)$ is a polynomial, exponential, sine, cosine function, or some linear combination of these functions.

The general solution of equations like (9.6) is perfectly standard as we have $y_g(x) = C\exp(-ax)$. It follows that the difficult part of the undetermined coefficients method lies in finding the particular solution. This usually involves an educated guess of the form of the solutions with "undetermined coefficients" (hence the name of the technique). In some cases, the form of the particular solution is reasonably obvious. In others, it may require quite a bit of work.

Most of the problems we solve using the undetermined coefficients method could also be solved by alternative methods such as the integrating factor method. In some cases, however, the solution is much easier using this approach, particularly if the integral of the right-hand side expression is difficult. The other big advantage of this approach is that it generalizes to higher-order differential equations. In the next chapter, we will see that the undetermined coefficients approach becomes the standard method when solving second-order differential equations.

EXAMPLE

Find the general solution of the differential equation $\frac{dy}{dx} + 2y = \exp(3x)$.

First, we note that the complementary function is easily found as $y_c(x) = C\exp(-2x)$. The difficult part here is finding the particular integral. In this case, the form of the expression on the right-hand side suggests an exponential function. Let us, therefore, try a function of the form $y_p(x) = A\exp(bx)$, where A and b are *undetermined coefficients*. Our task is now to determine these coefficients using the information given to us in the equation.

If $y_p(x) = A\exp(bx)$ is a solution, then our equation tells us that

$$bA\exp(bx) + 3A\exp(bx) = \exp(3x).$$

We can immediately see that $b = 3$. In fact, we could probably have safely assumed this from the start, but we left b as an unknown coefficient to illustrate the method. Setting $b = 3$ gives us

$$3A\exp(3x) + 2A\exp(3x) = \exp(3x).$$

This is true if $A = 1/5$. Therefore, the particular solution takes the form $y_p(x) = \exp(3x)/5$ and the general solution of the nonhomogeneous equation takes the form

$$y(x) = y_c(x) + y_p(x) = C\exp(-2x) + \frac{1}{5}\exp(3x).$$

In our second example, we assume that the function $f(x)$ is linear. As with the first example, this gives us a starting point for making an educated guess as to the form of the particular solution.

EXAMPLE

Find the general solution of the differential equation $\frac{dy}{dx}+\frac{1}{2}y=1+\frac{1}{4}x.$

As in the previous example, the complementary function here is perfectly standard. We have $y_c(x)=C\exp(-x/2)$. The difficult part is finding the particular integral. Given the linearity of the function on the right-hand side, we will assume a linear form for the particular integral, let $y_p(x)=a+bx$, where a and b are undetermined coefficients. From the differential equation, we have

$$b+\frac{1}{2}(a+bx)=1+\frac{1}{4}x.$$

Equating coefficients on the left and right-hand sides gives us b = 1/2 and a = 1. The particular integral takes the form $y_p(x)=1+x/2$ and the general solution to the nonhomogeneous equation is

$$y_g(x)=C\exp\left(-\frac{x}{2}\right)+1+\frac{1}{2}x.$$

So far, we have only applied the method of undetermined coefficients to cases in which the coefficient on the y variable is constant. It is possible to apply it in more general cases, but this becomes rather more difficult, and there is no guarantee that the method will be successful. However, to show how the method can be applied to a more general case, we will consider an example of its application to a problem in which the coefficient on y is a function of the independent variable x.

EXAMPLE

Find the general solution of the differential equation $\frac{dy}{dx}+2xy=3x.$

We note that the coefficient on y in this equation is equal to $2x$ and is not constant. This makes it more difficult to find the complementary function. However, we can do this by solving the associated homogeneous equation either by separation of variables or by the integrating factor method. Either of these approaches will yield the following solution.

$$y_c(x)=C\exp\left(-x^2\right).$$

For the particular integral, we note that the right-hand side is linear and choose a linear function of x as our initial guess. Let $y_p(x) = a + bx$, where a and b are undetermined coefficients. Substituting our guess into the differential equation gives us

$$b + 2x(a + bx) = 3x \text{ or } b + 2xa + 2bx^2 = 3x.$$

For this equation to be valid, we need $b = 0$ and $a = 3/2$. The particular integral; therefore, it takes a very simple form $y_p(x) = 3/2$, and the general solution for the original equation is

$$y_g(x) = y_c(x) + y_p(x) = C\exp(-x^2) + \frac{3}{2}.$$

REVIEW EXERCISES – SECTION 9.4

1. Find the general solution of the differential equation $\frac{dy}{dx} + 2y = 3x$ using the method of undetermined coefficients.
2. Find the general solution of the differential equation $\frac{dy}{dx} + y = \exp(2x)$ using the method of undetermined coefficients.
3. Find the particular solution of the differential equation $\frac{dy}{dx} + 0.5y = \exp(0.5x)$ with initial condition $y(0) = 10$ using the method of undetermined coefficients.

9.5 NUMERICAL METHODS

> Numerical methods can be used to solve differential equations when analytical solutions are not possible. They are also useful when we wish to apply theoretical models to real-world problems.

The simplest numerical method we can use to solve differential equations is *Euler's method*. Suppose we have a differential equation of the form $dy/dx = f(x,y)$, and we wish to solve this for the interval $x \in [a,b]$. The first step is to divide the interval up into n subintervals each of length h, where $h = (b-a)/n$. For given initial value, y_0, we can solve this using the recurrence relationship

$$
\begin{aligned}
x_{i+1} &= x_i + h \\
y_{i+1} &= y_i + hf(x_i, y_i) \\
i &= 0, \ldots, n-1.
\end{aligned}
$$

Figure 9.1 gives Python computer code for the calculation of the solution for a differential equation of the form $dy/dx = y$ with $y(0) = 1$. Note that this equation can be solved analytically to give the solution $y(x) = \exp(x)$. This will allow us to assess the accuracy of the numerical solution.

```
import numpy as np

def f(x,y):
    f = y
    return f

a = 0
b = 10
m = b-a
n = 10*m
h = (b-a)/n

x = np.zeros(n+1)
y = np.zeros(n+1)
y[0]=1

for idx in range(1,n+1):
    y[idx]=y[idx-1]+h*f(x[idx-1],y[idx-1])
    x[idx]=x[idx-1]+h

for idx in range(0,m+1):
    kdx = int(n*idx/m)
    print(idx,"\t","%0.1f" % x[kdx],"\t","%0.4f" % y[kdx])
```

FIGURE 9.1 Python code for Euler's method.

The problem with any numerical method for solving differential equations is that they are subject to error. In this case of Euler's method, errors arise because it uses a linear approximation to the function based on the differential $dy = f(x,y)dx$ in which we substitute a small interval h for dx. If the function is nonlinear, then this will inevitably result in an error. In the code given in Figure 9.1, we have set the interval $h = 1/10$. As the value of x increases, then the error will also increase. For $x = 10$, Euler's method gives a solution $y(10) = 13{,}780$, but we know that the true solution is $\exp(10) = 22{,}026$.

Therefore, the numerical solution underpredicts the true value by a factor of 37%. We can attempt to deal with this by shortening the interval h used in the recurrence relationship. If we set $h = 10^{-3}$, then Euler's method gives the solution $y(10) = 21{,}917$. This is closer to the true solution but requires 100 times as many function evaluations to calculate. Therefore, Euler's method is potentially expensive in terms of the number of calculations required to achieve an accurate solution.

An alternative approach is provided by the *Runge–Kutta method*[2]. Rather than relying on a simple linear approximation to the derivative of the function over the interval x to $x+h$, the Runge–Kutta method uses a weighted average of estimates of the slope based on the endpoints as well as two intermediate points for each interval. The recurrence relationship used to calculate the Runge–Kutta estimates takes the form.

$$
\begin{aligned}
&x_{i+1} = x_i + h\\
&k_1 = f(x_i, y_i)\\
&k_2 = f\left(x_i + \frac{h}{2}, y_i + h\frac{k_1}{2}\right)\\
&k_3 = f\left(x_i + \frac{h}{2}, y_i + h\frac{k_2}{2}\right)\\
&k_4 = f(x_i + h, y_i + hk_3)\\
&y_{i+1} = y_i + \frac{h}{6}(k_1 + 2k_2 + 2k_3 + k_4)\\
&i = 0, \ldots, n-1\,.
\end{aligned}
$$

Figure 9.2 shows Python code for the Runge–Kutta method. The effect of taking an average of multiple estimates of the gradient in each interval is to make the estimate used much more accurate. This means that we can set a much higher value of h and reduce the number of function evaluations while still achieving a higher level of accuracy. For example, in the case of the differential equation $dy/dx = y$ with $y(0) = 1$, if we set $h = 1/10$, then the Runge–Kutta method gives us an estimate of $y(10)$ equal to 22,026, which is accurate to one decimal place. To compare, the most accurate estimate

[2] Although we refer to the Runge-Kutta method, there exists a variety of similar algorithms which bear this name. The version discussed here is the most basic version which is known as the Runke–Kutta 4 (RK4) algorithm.

we obtained using Euler's method, with $h = 10^{-3}$, required 10,000 function evaluations. The Runge–Kutta method with $h = 0.1$, is both more accurate and only requires 400 function evaluations.

```
import numpy as np

def f(x,y):
    f = y
    return f

a = 0
b = 10
m = b-a
n = 10*m
h = (b-a)/n

x = np.zeros(n+1)
y = np.zeros(n+1)
y[0]=1

for idx in range(1,n+1):
    k1 = f(x[idx-1],y[idx-1])
    k2 = f(x[idx-1]+h/2,y[idx-1]+h*k1/2)
    k3 = f(x[idx-1]+h/2,y[idx-1]+h*k2/2)
    k4 = f(x[idx-1]+h,y[idx-1]+h*k3)
    y[idx]=y[idx-1]+h*(k1+2*k2+2*k3+k4)/6
    x[idx]=x[idx-1]+h

for idx in range(0,m+1):
    kdx = int(n*idx/m)
    print(idx,"\t","%0.1f" % x[kdx],"\t","%0.4f" % y[kdx])
```

FIGURE 9.2 Python code for the Runge–Kutta method.

Numerical methods can be used to check that explicit solutions of differential equations are correct. For example, in Section 9.2, we derived an explicit solution of the differential equation $dy/dx + y/x = 1$, which takes the form $y(x) = x/2 + C/x$. If we have initial condition $y(1) = 0$, then we can solve for the constant of integration as $C = -1/2$, which gives us an equation of the form $y(x) = x/2 - 1/(2x)$. We can check that this solution is correct by differentiating to show that we recover the original equation. However, we can also check by comparing the values obtained for $y(x)$, over a given range of values of x, from our solution and from a numerical solution obtained using the Runge–Kutta method. The results are shown in Table 9.1. As you can see, the results are identical to four decimal places. This indicates that the explicit solution we have obtained is correct.

TABLE 9.1 Comparison of the explicit solution of the differential equation $dy / dx + y / x = 1$ with the numerical solution obtained using the Runge–Kutta method.

x-Value	Explicit solution	Numerical Runge–Kutta solution
1	0.0000	0.0000
2	0.7500	0.7500
3	1.3333	1.3333
4	1.8750	1.8750
5	2.4000	2.4000
6	2.9167	2.9167
7	3.4286	3.4286
8	3.9375	3.9375
9	4.4444	4.4444
10	4.9500	4.9500

REVIEW EXERCISES – SECTION 9.5

1. Consider the differential equation $dy / dx = -0.2y$ where $y(0) = 1$. Using Euler's method, calculate estimates of $y(1)$ with intervals (a) $h = 0.5$, and (b) $h = 0.2$.

2. Using the computer code provided for the Runge–Kutta method, solve the differential equation $dy / dx = -0.5y$ with initial condition $y(0) = 100$ up to the value $x = 10$. Plot your solution on a graph with x values on the horizontal axis, and y values on the vertical axis.

9.6 SOME ECONOMIC EXAMPLES

In this section, we look at some of the ways in which differential equations can be used to model dynamic aspects of economics. We begin by looking at processes involving exponential growth and exponential decay. We then look at the more complex case of Cagan's model of inflation in which the differential equation to be solved is derived from a behavioral relationship.

The phrase *exponential growth* is often misused to describe situations in which a variable of interest grows extremely quickly. While this is sometimes true of exponential growth processes, it is not a defining feature. Variables that grow exponentially may exhibit quite modest changes over a certain time period. Similarly, a variable that grows very quickly may do so as the result of growth processes that are not exponential. In this context, the term exponential implies that the change in the value of the variable is proportional to its level. Thus, the differential equation describing a variable that grows exponentially can be written as

$$\frac{dy}{dt} = gy \, . \tag{9.7}$$

Note that we use t rather than x here to emphasize that this describes change through time. The general solution of an equation like this is easy to obtain by the method of separation of variables, and the solution takes the form

$$y(t) = Ae^{gt} \, , \tag{9.8}$$

where A is an arbitrary constant. To obtain the particular solution, we normally rely on some form of *boundary condition* to choose a particular value for A. For example, if the value of y is known for $t = 0$, then we use an *initial condition*. That is, if $y(0) = y_0$, then the particular solution takes the form $y(t) = y_0 e^{gt}$.

EXAMPLE

The level of real GDP for the United States can be modeled as an exponential growth process. The average growth rate between 1970 and 2019 was approximately 2.78% per annum, and the value of real GDP in 1970 was $4,954 billion at 2012 prices. An exponential growth model therefore takes the form $y(t) = 4{,}954\exp(t)$, where $t = 0$ in 1970 and increases by one in each successive year. The prediction for $t = 2019$ is therefore $y(49) = 4{,}954 \exp(0.0278 \times 49) = 19{,}346$. This is within 2% of the actual value of 19,033 where all figures are given in billions of dollars at 2012 prices.

In some cases, a *terminal condition* may be the more appropriate way to fix the value of the arbitrary constant in the general solution. This is often the case when modeling the value of financial assets. Consider, for example, a noninterest-bearing bond with a fixed date T at which it will be redeemed as some face value F. During the life of the bond, it must compete with alternative assets which bear interest at rate r. Hence, the value of the bond must increase through time at the rate r, and the differential equation, which describes the value of the bond at date t is given by $dV/dt = rV$, which has solution $V(t) = A\exp(rt)$. In this case, we use the terminal condition that $V(T) = F$ to determine the constant A. We have, $F = A\exp(rT)$, and we can write

$$V(t) = \frac{F}{\exp(rT)} = \exp(rt) = F\exp(r(t-T)), \tag{9.9}$$

as our solution.

EXAMPLE

A 10-year bond is issued with a face value of \$100. The market rate of interest is equal to 5%. What will be the value of the bond at date t, where $t \in [0,T]$?

Since the rate of interest is 5%, the value of the bond will be determined by the differential equation $dV/dt = 0.05V$, which has a general solution $V(t) = A\exp(0.05t)$. Since it will be redeemed at $t = 10$ for its face value of \$100, we have $100 = A\exp(0.5)$ which gives $A = 100/\exp(0.5) = 60.65$. In this case, we can write the particular solution in two equivalent ways. We have either $V(t) = 60.65\exp(0.05t)$ or $V(t) = 100\exp(0.05(t-10))$ for $t \in [0,10]$.

Models of exponential growth and decay are particularly important in economics. However, there are many situations in which economic models give rise to more general differential equations. As an example, we will consider Cagan's[3] (1956) model of inflation, which links the demand for real money balances to the rate of inflation. This model is particularly applicable to situations with very high rates of inflation (hyperinflation). The demand for real money balances in this model takes the form $m - p = -a\, dp/dt$ where m and p are the logarithms of the money stock and the price level, respectively. The money supply grows at rate σ so that $m(t) = \sigma t$. We can therefore write a differential equation for the determination of the price level, which takes the form

---

[3] Cagan, Phillip (1956). "The Monetary Dynamics of Hyperinflation". In Friedman, Milton (ed.). Studies in the Quantity Theory of Money. Chicago: University of Chicago Press. ISBN 0-226-26406-8.

$$\frac{dp}{dt}-\frac{1}{a}p=-\frac{1}{a}\sigma t. \tag{9.10}$$

This can be solved easily using the integrating factor method. We have

$$\frac{d\,pe^{-t/a}}{dt}=-\frac{\sigma}{a}e^{-t/a}t, \tag{9.11}$$

and integrating both sides yields:

$$pe^{-t/a}=-\frac{\sigma}{a}\int e^{-t/a}t\,dt. \tag{9.12}$$

The integral on the right-hand side is a standard integral of a form we covered in the previous chapter. We can write the solution to this equation as:

$$\begin{aligned} pe^{-t/a} &= e^{-t/a}\sigma(t+a)+C \\ \Rightarrow p(t) &= \sigma(t+a)+Ce^{t/a}. \end{aligned} \tag{9.13}$$

Now $Ce^{t/a}$ becomes arbitrarily large as $t\rightarrow\infty$ if $C\neq 0$. To avoid the price level becoming explosive, we require $C=0$. The solution for the logarithm of the price level takes the form $p(t)=\sigma(t+a)=m(t)+a\sigma$. Using capital letters to indicate the level of the variable rather than its logarithm, we have $P(t)=M(t)\exp(a\sigma)$. Thus, the level of the money supply determines the price level, but an increase in the rate of growth of money will produce a jump in prices even if there is no corresponding jump in the level of the money stock. This is because an increase in money growth increases the steady-state rate of inflation, which reduces the demand for money and requires an increase in the price level to maintain money market equilibrium. Therefore, an increase in money growth produces both a one-off increase in the price level and an increase in its growth rate.

The Solow (1956) growth model provides another example in which a differential equation arises naturally as part of an economic model. We assume that output per capita y is a function of a capital per capita k such that $y=k^{a}$, and that savings are a constant proportion of total output equal to s. In addition, we assume that the labor force grows at constant rate n and that capital decays at rate δ. It follows that the rate of change of the capital–labor ratio with respect to time is given by the differential equation

$$\frac{dk}{dt}=sk^{a}-(n+\delta)k. \tag{9.14}$$

This is quite a hard differential equation to solve analytically. However, we can say quite a bit about the nature of the solution. First, we note that there are two steady-states for this equation. The first occurs when $k=0$ and there is, therefore, no production, saving, or capital accumulation. However, this situation is unstable if there is even a tiny amount of capital in the initial state. If this is the case, then capital accumulation will begin, and the capital–labor ratio will converge on its other steady-state value in which $dk/dt=0$ and $k=\left[s/(n+\delta)\right]^{1/(1-a)}$. This is illustrated in Figure 9.3, which shows the relationship between dk/dt and k for the parameter values $s=0.2, a=0.25$, $n=0.025$, and $\delta=0.02$. This gives an equilibrium value for the capital–labor ratio equal to $k^*=7.3073$.

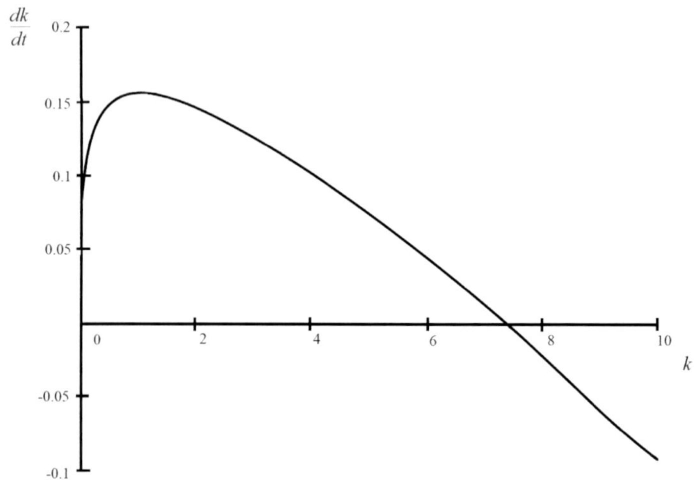

FIGURE 9.3 Relationship between capital accumulation and the capital-labor ratio in the Solow model.

From Figure 9.3, we see that when the initial capital–labor ratio is positive but lies below the steady-state value $dk/dt>0$, the system will trend toward a steady state. Similarly, if the capital–labor ratio lies above the steady-state value, then $dk/dt<0$, which again means that it will move toward the steady-state value. It follows that the value of $k>0$ at which $dk/dt=0$ is a stable equilibrium of the system.

Although equation (9.14) is hard to solve analytically, it is easy to solve numerically for given values of the parameters. In Table 9.2, we show the values of k and y at different points in time as calculated using the Runge–Kutta method, assuming the same parameter values used to construct Figure 9.3 and starting with $k(0)=5$. This illustrates the convergence of the system to equilibrium as the result of capital accumulation.

TABLE 9.2 Solution of Solow growth model by Runge-Kutta algorithm.

Time	*Capital–labor ratio*	*Output–labor ratio*
0	5.0000	1.4953
10	5.6383	1.5409
20	6.1053	1.5719
30	6.4440	1.5933
40	6.6885	1.6082
50	6.8644	1.6186
60	6.9905	1.6260
70	7.0809	1.6313
80	7.1456	1.6350
90	7.1918	1.6376
100	7.2248	1.6395
200	7.3045	1.6440
300	7.3072	1.6441

REVIEW EXERCISES - SECTION 9.6

1. Describe the effects of a cut in the money growth rate in the Cagan model of inflation.
2. Describe the effects of an increase in the rate of growth of the labor supply in the Solow growth model.

CHAPTER 10

SECOND-ORDER DIFFERENTIAL EQUATIONS

In this chapter, we show how to solve second-order differential equations and how the solutions can be interpreted. We will concentrate on linear differential equations because the solution of nonlinear equations is much more difficult for second-order equations than was the case for the first-order equations we considered in Chapter 9. This is because we cannot use the method of direct integration to solve equations of this type. There are, however, standard procedures for solving linear equations, which we will set out in this chapter. The general form of the equations we will consider is given in equation (10.1)

$$a_2 \frac{d^2 y}{dx^2} + a_1 \frac{dy}{dx} + a_0 y = f(x). \tag{10.1}$$

Equation (10.1) is a linear equation because the a coefficients do not depend on y. The special case $f(x) = 0$ is a *homogeneous* second-order differential equation and, if $f(x) \neq 0$, then we say that this is a *nonhomogeneous* equation. Equations of this type are often found in economic analysis as the result of the interaction in dynamic adjustment of related variables.

EXAMPLE

Suppose we have two dependent variables, y and z, which are linked through the following first-order differential equations.

$$\frac{dy}{dx} = -y + \frac{1}{2} z$$

$$\frac{dz}{dx} = \frac{1}{2} y - 2z.$$

We can express this system as a single second-order differential equation in terms of either y or z as follows. First, differentiate the first equation with respect to x to obtain

$$\frac{d^2y}{dx^2} = -\frac{dy}{dx} + \frac{1}{2}\frac{dz}{dx}.$$

Next, use the second equation to substitute for dz / dx to obtain

$$\frac{d^2y}{dx^2} = -\frac{dy}{dx} + \frac{1}{4}y - z.$$

Finally, solve the first equation for z to obtain $z = 2dy / dx + 2y$, substitute this into the transformed equation, and rearrange to get the final form of the equation.

$$\frac{d^2y}{dx^2} + 3\frac{dy}{dx} + \frac{7}{4}y = 0.$$

This is a general feature of all systems, which consist of a pair of first-order differential equations. Since many economic models give rise to systems of equations of this form, it is important to know how to solve them.

10.1 HOMOGENEOUS SECOND-ORDER LINEAR DIFFERENTIAL EQUATIONS

> The solution of homogeneous equations is an important first step in solving the more general problem of nonhomogeneous equations. In this section, we show how the general solution of homogeneous equations depends on the roots of its characteristic equation.

Let's start with a relatively easy case. Consider the equation

$$\frac{d^2y}{dx^2} + a_1\frac{dy}{dx} + a_0 y = 0. \tag{10.2}$$

This is a homogeneous equation with constant coefficients. If the equation had been nonhomogeneous, that is, if the right-hand side had not been equation to zero, then it would be much more difficult to solve. By starting with this case, we are making things much easier for ourselves. As we will see later, the solution of the homogeneous case forms part of the solution for nonhomogeneous equations, and, therefore, this is an important first step in the process of solving the more general case.

Now, when we solved first-order linear equations with constant coefficients, we found a general solution of the form $y_g(x) = C\exp(\lambda x)$, where λ is a parameter and C is a constant of integration which can be solved by using an initial condition. Would this solution work here? The question is whether or not we can find a value of λ which satisfies the differential equation. Differentiating our proposed solution gives us $dy_g / dx = \lambda C\exp(\lambda x)$ and $d^2y_g / dx^2 = \lambda^2 C\exp(\lambda x)$. Substituting into our differential equation gives us an expression of the form

$$C\exp(\lambda x)\left\{\lambda^2 + a_1\lambda + a_0\right\} = 0.$$

For this expression to be equal to zero for all possible values of the constant of integration C, we need the expression in the curly parentheses to be equal to zero. For a second-order differential equation, this expression is a quadratic function of the parameter λ, and we refer to this function as the *characteristic equation* for the problem. If we can find a value, or values of λ, which satisfy the equation $\lambda^2 + a_1\lambda + a_0 = 0$, then these will give us a solution, or solutions, to the differential equation.

This situation will often arise when solving second-order differential equations. Since the characteristic equation is quadratic, we will generally have two possible solutions. To choose the form of our general solution, we will make use of an important property of linear differential equations which is known as the *principle of superposition*. Let λ_1 and λ_2 be the solutions to the characteristic equation for the general problem (10.2). This means that we have possible solutions $y_1(x) = C_1\exp(\lambda_1 x)$ and $y_2(x) = C_2\exp(\lambda_2 x)$. The principle of superposition states that any linear combination of these solutions is itself also a solution. Since this principle is so important, we will state it formally below. There is a formal proof and extended discussion of this principle in the appendix.

Principle of Superposition

If $y_1(x)$ and $y_2(x)$ are solutions of a second-order linear differential equation, then so is $y(x) = k_1y_1(x) + k_2y_2(x)$, where k_1 and k_2 are constants.

EXAMPLE

Find the general solution of the differential equation $\frac{d^2y}{dx^2}+\frac{dy}{dx}-6y=0.$

The characteristic equation is given by $\lambda^2+\lambda-6=0$, which factorizes easily to give $(\lambda-2)(\lambda+3)=0$. There are therefore two roots $\lambda=2$ and $\lambda=-3$ and, by the principle of superposition, we can write the general solution of this equation as

$$y_g(x)=C_1\exp(3x)+C_2\exp(-2x)$$

where C_1 and C_2 are arbitrary constants of integration which will depend on initial conditions.

Our example illustrates an important property of the general solution for equations of this type. If the roots of the characteristic equation are real and distinct, as is the case for this example, then the solution can be written in the form

$$y_g(x)=C_1\exp(\lambda_1 x)+C_2\exp(\lambda_2 x).$$

Now, if both roots are negative then $y_g(x)$ tends to zero as $x\to\infty$. However, if either root is positive then the solution is explosive. For example, if $\lambda_1>0$ and $C_1\neq 0$ then the solution will either tend to ∞ if $C_1>0$, or $-\infty$ if $C_1<0$.

If the roots are complex conjugates then they will take the form $\lambda_1=\alpha+\beta i$ and $\lambda_2=\alpha-\beta i$, where α and β are real numbers and $i=\sqrt{-1}$. If this is the case, then we can still write the general solution of the differential equation in the form given by equation (10.3). However, there is a more convenient form that does not involve imaginary numbers. As we show in the appendix, the general solution when the roots are complex conjugates can be written in the form

$$y_g(x)=\exp(\alpha x)\{C_1\cos(\beta x)+C_2\sin(\beta x)\}.$$

Since both the sine and cosine functions are periodic, the expression in the curly parentheses will also be periodic. This solution will tend to zero if α is negative but will be explosive if α is positive.

EXAMPLE

Find the general solution of the differential equation $\frac{d^2y}{dx^2} - 2\frac{dy}{dx} + 5y = 0.$

The characteristic equation is $\lambda^2 - 2\lambda + 5 = 0$ which has roots $\lambda_1 = 1 + 2i$ and $\lambda_2 = 1 - 2i$. Using equation (10.4), we can write the general solution as

$$y_g(x) = \exp(x)\{C_1 \cos(2x) + C_2 \sin(2x)\}.$$

This solution is explosive because the real part of the roots is greater than zero.

If the roots are real but not distinct, then we have $a_1^2 = 4a_0$ and therefore $\lambda = -a_1/2$. For cases like this, the general solution can be shown to take the form

$$y_g(x) = C_1 \exp(\lambda x) + C_2 x \exp(\lambda x). \tag{10.5}$$

This result can be proved relatively straightforwardly by differentiating equation (10.5) twice to show that we recover the original equation. However, it is not presented here because it is somewhat lengthy and does not offer any significant insights.

EXAMPLE

Find the general solution of the differential equation $\frac{d^2y}{dx^2} + 6\frac{dy}{dx} + 9 = 0.$

The characteristic equation here is $\lambda^2 + 6\lambda + 9 = 0$ which factorizes to give $(\lambda + 3)^2 = 0$. It follows that the roots are real but not distinct, and we have $\lambda = -3$. Using equation (10.5), we can write the general solution as

$$y_g(x) = C_1 \exp(-3x) + C_2 x \exp(-3x).$$

In this case the general solution will converge to zero as x tends to infinity because the root is equal to minus three. In general, for cases of repeated roots, the condition for convergence remains the same as for distinct roots. If the root is negative, then $y_g(x) \to 0$ as $x \to \infty$. If, however, the root is positive, then $y_g(x)$ is explosive.

REVIEW EXERCISES – SECTION 10.1

1. Find the general solution of the differential equation $\frac{d^2y}{dx^2}+2\frac{dy}{dx}+\frac{5}{4}y=0$.

2. Find the general solution of the differential equation $\frac{d^2y}{dx^2}-10\frac{dy}{dx}+21y=0$.

3. Find the general solution of the differential equation $\frac{d^2y}{dx^2}-10\frac{dy}{dx}+25y=0$.

10.2 INITIAL VALUE PROBLEMS WITH SECOND-ORDER DIFFERENTIAL EQUATIONS

We need two initial conditions to solve for the particular solution of a second-order differential equation. In this section, we show how we can use initial conditions to solve for the constants of integration and demonstrate the types of solutions which can be found.

The principle of using initial conditions to eliminate the constants of integration is essentially the same for second-order differential equations as we saw earlier for first-order equations. The only real difference is that there are typically two unknown constants of integration in the general solution of the second-order equation, and we, therefore, need two initial conditions to obtain the particular solution. However, another qualitative difference between the solutions of first and second-order equations is that the latter can produce far more varied patterns of dynamic adjustment. In this section, we will use examples of second-order equations to illustrate the solution method and the types of dynamic paths which these can produce.

EXAMPLE 1: Distinct Real Roots

Consider the differential equation $\frac{d^2y}{dx^2}+3\frac{dy}{dx}+2y=0$ with initial conditions $y(0)=1$ and $y(-1)=0$.

To find the particular solution of this equation, we first find the general solution and then use the initial conditions to solve for the constants of integration. The first stage is to find the roots of the characteristic equation $\lambda^2+3\lambda+2=0$.

This factorizes easily to give us $(\lambda+1)(\lambda+2)=0$. We, therefore, have two real roots $\lambda_1=-1$ and $\lambda_2=-2$, both of which are negative. This means that the general solution can be written

$$y_g(x)=C_1\exp(-x)+C_2\exp(-2x).$$

From the initial conditions, we obtain a pair of simultaneous equations in the two unknown constants of integration. These are

$$C_1+C_2=1$$
$$C_1\exp(1)+C_2\exp(2)=0.$$

These can be solved to yield $C_1=1.5819$ and $C_2=-0.5819$. We can therefore write the particular solution for this problem, with the initial conditions given, as

$$y(x)=1.5819\exp(-x)-0.5819\exp(-2x).$$

We note that the fact that both roots are real and negative immediately tells us that the solution for this problem is convergent. That is $y(x)\to 0$ as $x\to\infty$. If we plot the solution we have obtained, as shown in Figure 10.1, then we confirm this property.

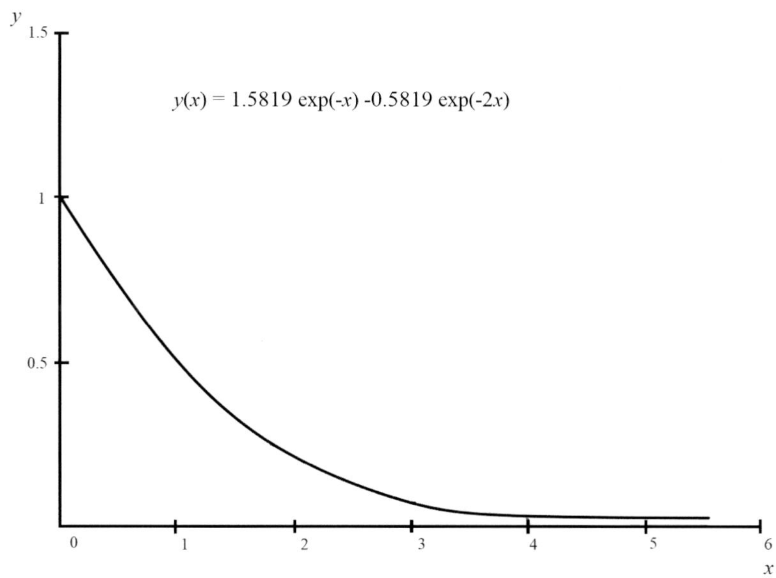

FIGURE 10.1 Solution path for second-order differential equation with negative real roots.

EXAMPLE 2: Complex Roots

Consider the differential equation $\frac{d^2y}{dx^2}+2\frac{dy}{dx}+10y=0$ with initial conditions $y(0)=0$ and $y(-1)=\exp(-1)$.

To find the particular solution to this equation, we again find the general solution and use the initial conditions to solve for the constants of integration. The first stage is to find the roots of the characteristic equation $\lambda^2+2\lambda+10=0$. This time, the factorization is more difficult, and we need to use the standard formula for quadratic equations to obtain

$$\lambda_{1,2}=\frac{-2\pm\sqrt{4-40}}{2}=-1\pm 3i.$$

Since the roots are complex conjugates, the general solution takes the form

$$y_g(x)=\exp(-x)\{C_1\cos(3x)+C_2\sin(3x)\}.$$

The initial conditions now give us the following pair of simultaneous equations

$$C_1\cos(0)=0$$
$$C_1\cos(-3)+C_2\sin(-3)=1.$$

These solve to give us $C_1=0$ and $C_2=-7.0862$. We can therefore write the particular solution for this problem, with the initial conditions given, as

$$y(x)=-7.0862\exp(-x)\sin(3x).$$

The fact that the roots have negative real components means that the solution will eventually converge to zero. In addition, the fact that they are complex conjugates means that we will observe cycles along the adjustment path. These properties can be seen in the solution illustrated in Figure 10.2.

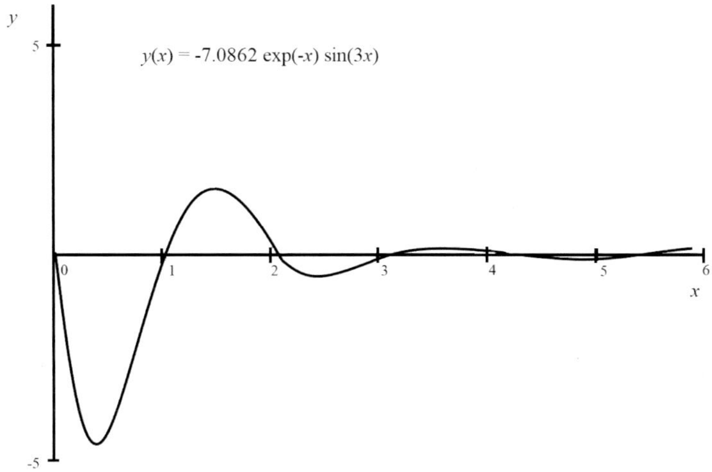

FIGURE 10. 2 Solution path for second-order differential equation with complex roots with a negative real part.

EXAMPLE 3: Repeated Real Roots

Consider the differential equation $\frac{d^2y}{dx^2} - 8\frac{dy}{dx} + 16y = 0$ with initial conditions $y(0) = 1$ and $\frac{dy}{dx}(0) = 0.$

Note that for this example, we have a different kind of initial condition in that, as well as fixing the value of y at $x = 0$, we also fix its derivative at this point. To find the particular solution, we first solve the characteristic equation $\lambda^2 - 8\lambda + 16 = 0$ to find the roots. This equation factorizes easily to give us $(\lambda - 4)^2 = 0$. There is a repeated root $\lambda = 4$ which is both real and positive. The general solution takes the form

$$y_g(x) = C_1 \exp(4x) + C_2 x \exp(4x).$$

The first derivative of this function is given by the expression

$$\frac{dy_g(x)}{dx} = 4C_1 \exp(4x) + C_2\{4x\exp(4x) + \exp(4x)\}.$$

Therefore, the initial conditions now give us the following pair of simultaneous equations

$$C_1 = 1$$
$$4C_1 + C_2 = 0$$

which solve to give us $C_1 = 1$ and $C_2 = -4$. We can therefore write the particular solution for this problem, with the initial conditions given, as

$$y(x) = \exp(4x) - 4x\exp(4x) = (1 - 4x)\exp(4x).$$

For $x > 1/4$ we have $y(x) < 0$ and, since the root is positive, this means that $y(x) \to -\infty$ as x becomes large.

In summary, we have shown how we can use initial conditions to solve for the constants of integration in second-order differential equations in exactly the same way as we did for first-order equations. However, we need two initial conditions when solving second-order equations. These can take the form of fixing the value of the solution at different points in time, but they can also take the form of fixing the value of the derivative of the function at some point. Second-order equations can generate more varied patterns of dynamic adjustment. Equations in which the roots are complex conjugates generate cyclical behavior. If the roots are real and either is positive, or complex, and have positive real roots, then the solution will exhibit explosive behavior. If the roots are real and negative, then the solution will approach zero smoothly. If the roots are complex with a negative real component, then the solution will tend to zero as x increases but will also exhibit cycles.

REVIEW EXERCISES – SECTION 10.2

Solve for the particular solution of each of the following differential equations using the initial conditions given

1. $8\frac{d^2y}{dx^2} + 6\frac{dy}{dx} + y = 0$ $\qquad y(0) = 2$

   $\left.\frac{dy}{dx}\right|_{x=0} = 0$

2. $\frac{d^2y}{dx^2} - 4\frac{dy}{dx} + \frac{17}{4}y = 0$ $\qquad y(0) = 1$

   $\left.\frac{dy}{dx}\right|_{x=0} = 0$

3. $9\frac{d^2y}{dx^2}+6\frac{dy}{dx}+y=0$ $\quad y(0)=3$

$y(-1)=0$

10.3 NONHOMOGENEOUS SECOND-ORDER LINEAR DIFFERENTIAL EQUATIONS

To solve nonhomogeneous differential equations, we make use of the superposition principle to divide the problem into two parts. First, we solve for the general solution of the related homogeneous equation, and then we add a particular integral to form the general solution of the nonhomogeneous equation.

In the previous sections, we have shown how to solve homogeneous second-order linear differential equations. This is an essential building block in the process of developing a method for the more general case of nonhomogeneous equations. Consider the following equation

$$\frac{d^2y}{dx^2}+a_1(x)\frac{dy}{dx}+a_0(x)y=f(x).$$

This defines the general case of a nonhomogeneous second-order linear differential equation. The functions $a_1(x)$, $a_0(x)$, and $f(x)$ are assumed to be continuous and integrable. In this section, we show how we can extend the methods we have developed for nonhomogeneous equations to solve equations of this type.

Our strategy for solving second-order nonhomogeneous equations is similar to that which we used for the first-order case. Let $y_c(x)$ be the general solution of the associated homogeneous model with $f(x)=0$, and let $y_p(x)$ be any particular integral of the nonhomogeneous equation. By the principle of superposition, $y_g(x)=y_c(x)+y_p(x)$ is the general solution of the nonhomogeneous equation. The procedure for finding the general solution of the homogeneous equation is well established and so, in practice, the more difficult part here is finding a particular integral. In most cases, we rely on making an educated guess as to the form of the solution and then using the method of undetermined coefficients to choose the specific parameters.

EXAMPLE

Find the general solution of the nonhomogeneous differential equation $\frac{d^2y}{dx^2}+3\frac{dy}{dx}+2y=3x.$

The general solution of the homogeneous model is straightforward. The characteristic equation is $\lambda^2+3\lambda+2=0$ which factorizes to give $(\lambda+2)(\lambda+1)=0$ and the general solution therefore takes the form

$$y_c(x)=C_1\exp(-x)+C_2\exp(-2x).$$

This now acts as the complementary function for the nonhomogeneous case. To find a particular integral, we will start with a guess as to the functional form. Since the expression on the right-hand side is a linear function of x, we will assume a linear function of the form $y_p(x)=a+bx$. Applying the method of undetermined coefficients, we have

$$3b+2(a+bx)=3x.$$

Equating coefficients allows us to solve for the parameter values as $b=3/2$ and $a=-9/4$. The particular integral, therefore, takes the form $y_p(x)=-9/4+(3/2)x$ which means that we can write the general solution of the nonhomogeneous equation as

$$y_g(x)=C_1\exp(-x)+C_2\exp(-2x)-\frac{9}{4}+\frac{3}{2}x.$$

This method relies on us being able to determine the correct functional form for the particular integral. There is no definitive way of doing this, but, in general, we can use the functional form of the driving function $f(x)$ as a guide. In the case of models with constant coefficients, this method will generally be reliable. Let us consider an alternative example to illustrate this.

EXAMPLE

Find the general solution of the nonhomogeneous differential equation $\frac{d^2y}{dx^2}+\frac{dy}{dx}-6y=4\exp(-x).$

As with the previous example, the complementary function is easy to derive because the characteristic polynomial factorizes to give us roots $\lambda_1=-3$ and

$\lambda_2 = 2$. We can, therefore, immediately write down the complementary function as

$$y_c(x) = C_1 \exp(-3x) + C_2 \exp(2x).$$

To determine the particular integral, we note that the right-hand side of our equation is an exponential function. We, therefore, choose an exponential functional form with general parameters A and b, that is, $y_p = A\exp(bx)$. Equating coefficients now gives us

$$b^2 A\exp(bx) + bA\exp(bx) - 6A\exp(bx) = 4\exp(-x).$$

It is immediately obvious that the only possible solution is one in which $b = -1$. We can therefore solve for A by substituting this value and writing our equation as

$$\exp(-x)\{A - A - 6A\} = 4\exp(-x).$$

It follows that $-6A = 4$ or $A = -2/3$. We can therefore write the general solution of the nonhomogeneous equation as

$$y_g(x) = C_1 \exp(-3x) + C_2 \exp(2x) - \frac{2}{3}\exp(-x).$$

Finally, we note that solving for a particular solution, which is consistent with given initial conditions, does not create any new problems in the case of nonhomogeneous equations. As in the case of a homogeneous second-order equation, we will need two boundary conditions to determine the two constants of integration C_1 and C_2. The procedure is exactly the same as we discussed in the previous section. To see this, let us consider one further example.

EXAMPLE

Find the particular solution of the differential equation $\frac{d^2y}{dx^2} - 3y = x^2$ with initial conditions $y(0) = 1$ and $\left.\frac{dy}{dx}\right|_{x=0} = 0$.

The characteristic equation here takes the form $\lambda^2 - 3 = 0$, and therefore, the roots are $\lambda_1 = \sqrt{3}$ and $\lambda_2 = -\sqrt{3}$. The complementary function, therefore, takes the form

$$y_c(x) = C_1 \exp\left(\sqrt{3}x\right) + C_2 \exp\left(-\sqrt{3}x\right).$$

Since the right-hand side of our equation is quadratic, let us try a general quadratic function for our particular integral. Let $y_p(x) = a + bx + cx^2$, where a, b, and c are unknown parameters. Equating coefficients gives us $2c - 3\left(a + bx + cx^2\right) = x^2$ which we can solve to give $b = 0$, $c = -1/3$, and $a = -2/9$. The general solution of the nonhomogeneous problem, therefore, takes the form

$$y_g(x) = C_1 \exp\left(\sqrt{3}x\right) + C_2 \exp\left(-\sqrt{3}x\right) - \frac{2}{9} - \frac{1}{3}x^2.$$

From the initial conditions, we have

$$C_1 + C_2 - \frac{2}{9} = 1$$

$$\sqrt{3}C_1 - \sqrt{3}C_2 - \frac{2}{3} = 0.$$

These equations solve to give $C_1 = 0.8036$ and $C_2 = 0.4187$. Therefore, the particular solution, which is consistent with these initial conditions, is given by the equation

$$y(x) = 0.8036 \exp\left(\sqrt{3}x\right) + 0.4187 \exp\left(-\sqrt{3}x\right) - \frac{2}{9} - \frac{1}{3}x^2.$$

REVIEW EXERCISES – SECTION 10.3

1. Consider the differential equation $d^2y/dx^2 + 3dy/dx + 2y = f(x)$. Find the complementary function and then, for each of the following functions $f(x)$, calculate a particular integral. Hence, find the general solution in each case.

(a) $f(x) = 2 + 3x$

(b) $f(x) = 4x^2$

(c) $f(x) = 2\exp\left(\frac{x}{2}\right)$

2. Find the particular solution of the differential equation $d^2y/dx^2 + dy/dx - 12y = 2x$ with initial conditions $y(0) = -1/72$ and $\left.\frac{dy}{dx}\right|_{x=0} = 1$.

10.4 NUMERICAL SOLUTION FOR SECOND-ORDER EQUATIONS

> Numerical methods offer a method of solving problems when analytical methods become either too difficult or, in some cases, impossible. These methods become essential when applying differential equations to complex real-world problems.

To solve second-order linear differential equations numerically, we note that any second-order linear equation can be written as a pair of linked first-order equations by making an appropriate substitution. For example, suppose we have an equation of the form

$$\frac{d^2y}{dx^2} + a_1(x)\frac{dy}{dx} + a_0(x)y = f(x).$$

If we define $z = dy/dx$, then we can write this in the form

$$\frac{dz}{dx} = f(x) - a_1(x)z - a_0(x)y = f_1(x, y, z)$$
$$\frac{dy}{dx} = z = f_2(x, y, z).$$

The functional form f_2 has deliberately been kept very general here, even though, for this particular case, dy/dx depends on z only. This is so the updating formulas, which we will now set out, will continue to be valid for more general cases. Using this notation, we can now set out updating formulas for the Runge–Kutta method as shown below:

$$k_{11} = f_1\left(x_k, y_k, z_k\right)$$

$$k_{21} = f_2\left(x_k, y_k, z_k\right)$$

$$k_{12} = f_1\left(x_k + \frac{h}{2}, y_k + \frac{h}{2}k_{11}, z_k + \frac{h}{2}k_{21}\right)$$

$$k_{22} = f_2\left(x_k + \frac{h}{2}, y_k + \frac{h}{2}k_{11}, z_k + \frac{h}{2}k_{21}\right)$$

$$k_{13} = f_1\left(x_k + \frac{h}{2}, y_k + \frac{h}{2}k_{12}, z_k + \frac{h}{2}k_{22}\right)$$

$$k_{23} = f_2\left(x_k + \frac{h}{2}, y_k + \frac{h}{2}k_{12}, z_k + \frac{h}{2}k_{22}\right)$$

$$k_{14} = f_1\left(x_k + h, y_k + hk_{13}, z_k + hk_{23}\right)$$

$$k_{24} = f_2\left(x_k + h, y_k + hk_{13}, z_k + hk_{23}\right)$$

$$z_{k+1} = z_k + \frac{h}{6}(k_{11} + 2k_{12} + 2k_{13} + k_{14})$$

$$y_{k+1} = y_k + \frac{h}{6}(k_{21} + 2k_{22} + 2k_{23} + k_{24})$$

$$x_{k+1} = x_k + h$$

EXAMPLE

Suppose we wish to solve the differential equation $\frac{d^2y}{dx^2} - \frac{dy}{dx} - 2y = 1 + 3x$ with initial conditions $y(0) = 10$ and $\left.\frac{dy}{dx}\right|_{x=0} = 0$.

This equation can be solved analytically to obtain the following expression

$$y(x) = \frac{15}{4}\exp(2x) + 6\exp(-x) + \frac{1}{4} - \frac{3}{2}x. \tag{10.6}$$

We can use this equation to calculate exact values of y for given values of x and compare these with approximate numerical solutions calculated using either the Euler or the Runge–Kutta method. Note that the presence of a positive root in the characteristic polynomial means that the solution will be explosive.

TABLE 10.1 Python code for Runge–Kutta solution for equation $\frac{d^2y}{dx} - \frac{dy}{dx} - 2y = 1 + 3x$ with initial conditions $y(0) = 10$ and $dy/dx|_{x=0} = 0$.

```
import numpy as np
import math

def f1(x,y1,y2):
    f1 = y2+2*y1+1+3*x
    return f1

def f2(x,y1,y2):
    f2=1*y2
    return f2

a = 0
b = 10
m = b-a
n = 100*m
h = (b-a)/n

y1 = np.zeros(n+1)
y2 = np.zeros(n+1)
x = np.zeros(n+1)

x[0]=0
y1[0]=10
y2[0]=0

for idx in range(1,n+1):
    k11 = f1(x[idx-1],y1[idx-1],y2[idx-1])
    k21 = f2(x[idx-1],y1[idx-1],y2[idx-1])
    k12 = f1(x[idx-1]+h/2,y1[idx-1]+h*k11/2,y2[idx-1]+h*k21/2)
    k22 = f2(x[idx-1]+h/2,y1[idx-1]+h*k11/2,y2[idx-1]+h*k21/2)
    k13 = f1(x[idx-1]+h/2,y1[idx-1]+h*k12/2,y2[idx-1]+h*k22/2)
    k23 = f2(x[idx-1]+h/2,y1[idx-1]+h*k12/2,y2[idx-1]+h*k22/2)
    k14 = f1(x[idx-1]+h,y1[idx-1]+h*k13,y2[idx-1]+h*k23)
    k24 = f2(x[idx-1]+h,y1[idx-1]+h*k13,y2[idx-1]+h*k23)
    y2[idx]=y2[idx-1]+h*(k11+2*k12+2*k13+k14)/6
    y1[idx]=y1[idx-1]+h*(k21+2*k22+2*k23+k24)/6
    x[idx]=x[idx-1]+h

print("Period","\t","Solution")
print()

for idx in range(0,m+1):
    kdx = int(n*idx/m)
    ranum = "%6.6f" % y1  [kdx]
    ranum = ranum.rjust(10)
    print(idx,"\t",ranum)
```

Table 10.2 compares the exact solution for $x = 1,...,5$ with the numerical solutions obtained using the Python code given in Table 10.1. From Table 10.2, we see that the Euler and Runge–Kutta solutions are roughly comparable in terms of their accuracy. The Euler solution is slightly closer to the exact solution for $x = 1$, but for all other values, the Runge–Kutta solution is more accurate. The difference, however, is that we set $h = 0.001$ for the Euler solution and $h = 0.01$ for the Runge–Kutta. This drastically reduces the number of function evaluations needed. For these calculations, the Euler method required 20,000 function evaluations, whereas the Runge–Kutta required only 8,000. With modern computing speeds, this made very little difference. However, for more complex problems requiring a higher degree of accuracy, the superior efficiency of the Runge–Kutta method might well become important.

TABLE 10.2 Exact and numerical solutions for second-order differential equation.

x	*Exact solution*	*Solution using Euler's method* h = 0.001	*Solution using Runge–Kutta method* h = 0.01	*% Error Euler's method*	*% Error Runge–Kutta method*
1	28.666237	28.609844	28.551336	–0.20	–0.40
2	202.80507	201.98801	202.10039	–0.40	–0.35
3	1508.9067	1499.8683	1503.8400	–0.60	–0.34
4	11172.952	11083.998	11135.617	–0.80	–0.33
5	82592.037	81771.250	82316.239	–0.99	–0.33

REVIEW EXERCISES – SECTION 10.4

1. Show that each of the following second-order differential equations can be represented as a pair of linked first-order equations.

(a) $\frac{d^2y}{dx^2} + 3x\frac{dy}{dx} + 2y = x$

(b) $4x\frac{d^2y}{dx^2} - 2y = \exp(x)$

2. Using the code provided, solve the equation given in part (b) using the Runge–Kutta method for values of x in the range 1 to 10, with initial conditions $y(1) = 1$ and $dy/dx\big|_{x=1} = 0$.

APPENDIX: THE PRINCIPLE OF SUPERPOSITION

The principle of superposition is particularly important for the solution of second-order linear differential equations. It can be stated as follows.

Let $y_1(x)$ be a solution of the second-order linear differential equation

$$\frac{d^2y}{dx^2}+a_1(x)\frac{dy}{dx}+a_0(x)y=f_1(x).$$

Note that this equation is linear in y and its derivatives, but there is no requirement for functions $a_1(x)$, $a_0(x)$, and $f_1(x)$ to be linear. All that is required is that these functions are continuous and integrable. Next, let $y_2(x)$ be a solution of the equation

$$\frac{d^2y}{dx^2}+a_1(x)\frac{dy}{dx}+a_0(x)y=f_2(x).$$

These equations differ only in the forcing function f on the right-hand side. The principle of superposition states that, for any constants k_1 and k_2, the function $k_1y_1(x)+k_2y_2(x)$ is a solution of the differential equation

$$\frac{d^2y}{dx^2}+a_1(x)\frac{dy}{dx}+a_0(x)y=k_1f_1(x)+k_2f_2(x).$$

Proof: Differentiating the weighted average function $k_1y_1(x)+k_2y_2(x)$ means that we can write the differential equation as

$$\begin{aligned}
&\frac{d^2\left(k_1y_1(x)+k_2y_2(x)\right)}{dx^2}+a_1(x)\frac{d\left(k_1y_1(x)+k_2y_2(x)\right)}{dx}+a_0(x)\left(k_1y_1(x)+k_2y_2(x)\right)\\
&=k_1\left(\frac{d^2\left(y_1(x)\right)}{dx^2}+a_1(x)\frac{dy_1(x)}{dx}+a_0(x)y_1(x)\right)\\
&\quad+k_2\left(\frac{d^2\left(y_2(x)\right)}{dx^2}+a_1(x)\frac{dy_2(x)}{dx}+a_0(x)y_2(x)\right)\\
&=k_1f_1(x)+k_2f_2(x).
\end{aligned}$$

This result proves to be important in a number of different contexts.

1. When solving for the general solution of a second-order differential equation with constant coefficients, we get a pair of solutions of the form $y(x) = C\exp(\lambda_i x);\ i = 1,2,$ where λ_i are the roots of the characteristic polynomial. The principle of superposition establishes that a weighted average of these solutions is also a solution.
2. When solving any nonhomogeneous linear differential equation, the principle of superposition establishes that the general solution is given by the sum of the complementary function (the general solution of the corresponding homogeneous equation) and a particular integral.

Note that, although we have presented the proof of the principle of superposition in terms of a second-order linear differential equation, this can easily be extended to any order of the differential equation. It can therefore be applied to the case of a first-order equation and used to demonstrate that the general solution of a nonhomogeneous equation is equal to the sum of the complementary function and a particular integral. It can also be extended to apply to higher-order differential equations. The only requirements are that the equation is linear in y and its derivatives, and that the coefficient functions and forcing function are continuous and integrable.

APPENDIX: DERIVATION OF THE COMPLEMENTARY FUNCTION WHEN THE ROOTS ARE COMPLEX

If the roots are complex, such that $\lambda_1 = \alpha + \beta i$ and $\lambda_2 = \alpha - \beta i$, then we can still write down solutions of the form

$$y_1(x) = \exp\{(\alpha + \beta i)x\}$$
$$y_2(x) = \exp\{(\alpha - \beta i)x\}.$$

Euler's formula allows us to write

$$\exp((\alpha + \beta i)x) = \exp(\alpha x)(\cos(\beta x) + i\sin(\beta x))$$
$$\exp((\alpha - \beta i)x) = \exp(\alpha x)(\cos(\beta x) - i\sin(\beta x)).$$

By the principle of superposition, we can define

$$u(x)=\frac{1}{2}(y_1(x)+y_2(x))=\exp(\alpha x)\cos(\beta x)$$
$$v(x)=\frac{1}{2i}(y_1(x)-y_2(x))=\exp(\alpha x)\sin(\beta x)$$

which are both real valued functions. Again, by the principle of superposition, we can take a weighted average of these two functions which gives us the complementary function

$$y_c(x)=\exp(\alpha x)\{C_1\cos(\beta x)+C_2\sin(\beta x)\}$$

where C_1 and C_2 are constants of integration which can be determined using boundary conditions.

CHAPTER 11

DIFFERENCE EQUATIONS

Difference equations are closely related to differential equations. Differential equations model continuous changes, whereas difference equations model discrete changes of one variable in response to others. Difference equations arise frequently in economics when modeling changes over time. In this chapter, we develop general methods for solving these types of equations and illustrate them with examples drawn from economic theory.

11.1 FIRST-ORDER DIFFERENCE EQUATIONS

In this section, we consider linear first-order difference equations. The solution method for equations of this type is very similar to that for linear first-order differential equations.

Consider an equation of the form

$$y_n - ay_{n-1} = f(n). \tag{11.1}$$

This is the general form of a linear first-order difference variable in y with a constant coefficient a. The variable y is observed at discrete intervals which are indexed by the subscript n. In most problems of this type, n takes on integer values only. This form of the function is *nonhomogeneous* if $f(n) \neq 0$. The solution method for equations of this type is based on the principle of superposition that we used when solving first-order linear differential equations.

Let $\tilde{y}_n$ be a general solution of the homogeneous equation obtained by setting $f(n)=0$ in equation (11.1). It is easy to see that, if $y_n - ay_{n-1} = 0$, then the general solution can be written as $\tilde{y}_n = Ca^n$ where C is a constant. This can easily be demonstrated by substituting into our equation to obtain $Ca^n - aCa^{n-1} = 0$ which is clearly true for all values of the arbitrary constant C.

Now, let us turn to the nonhomogeneous part of the equations. Let $\hat{y}_n$ be a particular solution of the nonhomogeneous equation, that is an equation of the form $\hat{y}(n) = g(n)$ which satisfies the nonhomogeneous equation and does not include any arbitrary constants. In most cases, we make an educated guess about the form of the function $g(n)$, using the form of the function $f(n)$ as a guide, and then use the method of undetermined coefficients to find parameter values that make it consistent with the equation of interest. Since our equation is linear, the principle of superposition tells us that $\tilde{y}_n + \hat{y}_n$ will be a general solution of the nonhomogeneous equation. Therefore, to solve an equation of the form (11.1), we use a similar strategy to that which we used for a first-order linear differential equation. First, we find a general solution to the homogeneous equation associated with our problem of interest. Next, we find a particular solution for the nonhomogeneous problem. Finally, we take the sum of our two solutions as the general solution for the nonhomogeneous case.

EXAMPLE

Find the general solution of the first-order linear differential equation $y_n = \frac{1}{2} y_{n-1} + 1$.

First, we can immediately write down the general solution of the associated homogeneous equation as $\tilde{y}_n = C(1/2)^n$. To get the second part of our solution, we need to find a particular solution (or particular integral) of the nonhomogeneous equation. Since the nonhomogeneous part of the equation of interest consists of a constant term, let us try a solution of the form $y_p = c$. Next, we use the method of undetermined coefficients to find a value for c. Substituting $y_p = c$ into our equation gives us $c = c/2 + 1$ or $c = 2$ and, therefore, $y_p = 2$ is a particular solution. Combining our two solutions gives us the general solution of the nonhomogeneous equation, which takes the form

$$y_n = \tilde{y}_n + y_p = C\left(\frac{1}{2}\right)^n + 2.$$

Note that we can easily check that this solution is correct by substituting it back into our original equation to show that it is consistent.

The example above generalizes to any first-order linear difference equation, a constant coefficient a and a constant intercept a_0. Consider the general equation

$$y_n = a_1 y_{n-1} + a_0 \tag{11.2}$$

where a_1 and a_0 are constants. The general solution of the associated homogeneous problem takes the form $\tilde{y}_n = Ca_1^n$, and it is straightforward to show that there is a particular solution $y_p = a_0 / (1 - a_1)$. It follows that the general solution for the nonhomogeneous problem takes the form

$$y_n = Ca_1^n + \frac{a_0}{1 - a_1}. \tag{11.3}$$

EXAMPLE

Find the general solution of the nonhomogeneous difference equation $y_n = 2y_{n-1} + 1$.

Using the general form given in equation (11.3) we can immediately write down the general solution as

$$y_n = C(2)^n + \frac{1}{1-2} = C(2)^n - 1.$$

From the general formula given in equation (11.3), we note that, if $|a_1| < 1$, then $Ca_1^n \to 0$ as $n \to \infty$ and we can regard the particular solution $a_0 / (1 - a_1)$ as the equilibrium value of y. However, if this condition is not satisfied, then the solution does not converge. This is the case for our example here in which $a = 2$ and therefore $Ca^n \to \infty$ as $n \to \infty$ unless $C = 0$.

The general solution for a difference equation includes an arbitrary constant of integration C. As in the case of differential equations, we will need an initial or boundary condition to eliminate this constant to solve for a particular solution of the nonhomogeneous equation. An initial condition consists of a specific value for y when $n = 0$, which will allow us to solve for C as demonstrated in the following example.

EXAMPLE

Find the particular solution of the nonhomogeneous difference equation $y_n = 0.25y_{n-1} + 4$ with initial condition $y_0 = 2$.

The general solution for this equation is the sum of the general solution of the associated homogeneous equation and a particular integral. We have

$$y_n = C(0.25)^n + \frac{4}{1-0.25} = C(0.25)^n + \frac{16}{3}.$$

From our initial condition, we have $2 = C + 16/3$ which solves to give us $C = -10/3$. The particular solution of the nonhomogeneous equation which is consistent with the initial condition is therefore

$$y_n = -\frac{10}{3}(0.25)^n + \frac{16}{3}.$$

First-order difference equations arise in dynamic economic models where variables of interest adjust over time. This is often the result of costs of adjustment which prevent agents from immediately adjusting choice variables to equilibrium values following a change in exogenous factors. For example, consider a macroeconomic model in which imports (m) depend on national income (y). If income changes, importers may not immediately change their demand levels for a variety of reasons including costs of adjustment. In the following example, we will show how we can model import demand using a difference equation and how we can solve this to determine the level of imports following a change in national income.

EXAMPLE

The demand for imports in an economy is determined by the difference equation $m_t = 0.5m_{t-1} + 0.2y$, where y is national income[1]. Now let $y = 1{,}000$ and $m_0 = 300$. Solve for the time path of imports.

The general solution of our difference equation for imports takes the form

$$m_t = C(0.5)^t + \frac{0.2 \times 1{,}000}{1-0.5} = C(0.5)^t + 400.$$

---

[1] Note the switch to t as the subscript here since the problem is explicitly one of adjustment over time. This is often, but not always, the case when using difference equations. For most of the text we will use the more general subscript n but we will switch to t in cases where this is appropriate.

From the initial condition, we have $300 = C + 400$ or $C = -100$. Therefore, the particular solution, which gives us the time path of imports, is given by the equation

$$m_t = -100(0.5)^t + 400.$$

Note that, in the long run as $t \to \infty$, the level of imports will converge on its equilibrium value of 400. Solutions of this type, in which the variable of interest converges on a constant, are referred to as stable solutions.

So far, we have only considered cases in which the nonhomogeneous part of the equation of interest takes the form of a constant. We can, however, use this method to solve more general difference equations in which the nonhomogeneous part of the equation is a function of n, provided we can find a suitable particular integral. The procedure parallels that of finding a particular integral in the case of differential equations. To see how this works, let us consider an example.

EXAMPLE

Find the particular solution of the nonhomogeneous difference equation $y_n = \left(\frac{1}{3}\right) y_{n-1} + 2n$ with initial condition $y_0 = 1$.

The solution of the associated homogeneous equation is obvious, and we can immediately write it as $\tilde{y}_n = C(1/3)^n$. The only novelty here lies in the solution for the particular integral. Since the nonhomogeneous part of our equation consists of a linear function of n, let us assume a linear particular integral of the form $y_p = a + bn$, where a and b are unknown parameters, and use the method of undetermined coefficients to find their values. From our difference equation, we have

$$a + bn = \left(\frac{1}{3}\right)\left(a + b(n-1)\right) + 2n$$

$$\left(\frac{2}{3}a + \frac{1}{3}b\right) + \left(\frac{2}{3}b\right)n = 2n.$$

Equating coefficients gives us solutions $b = 3$ and $a = -3/2$ and therefore, the general solution of the nonhomogeneous equation is

$$y_n = C\left(\frac{1}{3}\right)^n - \frac{3}{2} + 3n.$$

From our initial condition, we have $1 = C - 3/2$ or $C = 5/2$. Therefore, the particular solution which is consistent with the initial condition is given by

$$y_n = \frac{5}{2}\left(\frac{1}{3}\right)^n - \frac{3}{2} + 3n.$$

REVIEW EXERCISES – SECTION 11.1

1. Find the general solutions for the following difference equations. In each case, giving reasons, state whether y converges on the particular solution as $n \to \infty$.

(a) $y_n = 2y_{n-1} + 4$

(b) $y_n = -\frac{1}{2}y_{n-1} + 2$

(c) $y_n = -3y_{n-1} + 1$

2. Find the particular solutions of the following difference equations using the initial conditions given.

(a) $y_n = \frac{1}{5}y_{n-1} - 10 \qquad y_0 = 1$

(b) $y_n = -\frac{1}{5}y_{n-1} + 10 \qquad y_0 = 1$

3. Find the general solution of the following difference equation

$$y_n = \frac{1}{4}y_{n-1} + \exp\left(-\frac{1}{2}n\right).$$

11.2 SECOND-ORDER DIFFERENCE EQUATIONS

Second-order difference equations include two lags of the variable of interest. This will make them more difficult to solve than first-order equations. However, the solution method remains essentially the same.

The general form of a nonhomogeneous second-order linear difference equation with constant coefficients can be written as

$$y_n = a_1 y_{n-1} + a_2 y_{n-2} + f(n). \qquad (11.4)$$

We will again be looking for solutions to equations of this type which take the form $y_n = g(n)$. The solution method is essentially the same as for first-order equations. To find the general solution of the nonhomogeneous equation, we first look for a general solution to the associated homogeneous problem and, then for a particular solution of the nonhomogeneous problem. By the principle of superposition, the sum of these two solutions gives us a general solution for the nonhomogeneous problem. In the case of second-order equations, this will include two arbitrary constants of integration. To obtain a particular solution, we, therefore, need two initial, or boundary, conditions.

We will begin with the general solution for the homogeneous problem. Consider the equation

$$y_n - a_1 y_{n-1} - a_2 y_{n-2} = 0.$$

Since a solution of the form $y_n = C\lambda^n$ worked for the first-order case, let us try it for this case and see if we can find a value, or values for λ which will work for the second-order problem. Substituting our proposed solution into the equation gives us

$$C\lambda^n - a_1 C\lambda^{n-1} - a_2 C\lambda^{n-2} = 0.$$

Assuming C and λ are not zero, we can divide this expression by $C\lambda^{n-2}$ to obtain the equation

$$\lambda^2 - a_1\lambda - a_2 = 0$$

This is referred to as the *characteristic equation* for the second-order problem. The roots of this equation give us the values of λ which are consistent with $y_n = C\lambda^n$ being a solution of the homogeneous difference equation. Given that the characteristic equation is quadratic, there are three possible cases of interest.

Case 1: Real Distinct Roots

If $a_1^2 + 4a_2 > 0$, then the roots of the characteristic equation are real and distinct. This means that we have two possible general solutions $y_{1,n} = C_1\lambda_1^n$ and $y_{2,n} = C_2\lambda_2^n$. By the principle of superposition, it follows that the sum of these solutions will also be a solution and we can write down a general solution of the equation as

$$y_n = C_1\lambda_1^n + C_2\lambda_2^n \tag{11.5}$$

where C_1 and C_2 are arbitrary constants of integration. Note that it follows immediately that we need $|\lambda_1| < 1$ and $|\lambda_2| < 1$ for $y_n \to 0$ as $n \to \infty$. If either of the roots is greater than one in absolute value, then the solution will be explosive.

Case 2: Complex Roots

If $a_1^2 + 4a_2 < 0$, then the roots are complex conjugates of the form $\lambda_{1,2} = c \pm di$ where $c = a_1 / 2$ and $d = \left(\sqrt{4a_2 + a_1^2}\right)/2$. Now, if this is the case, we can still write the solution in the form $y_n = A_1\lambda_1^n + A_2\lambda_2^n$, where A_1 and A_2 are complex conjugates, but this is not particularly helpful. A more convenient form for the solution is

$$y_n = r^n\left(C_1\cos(\theta n) + C_2\sin(\theta n)\right)$$

where $r = \sqrt{c^2 + d^2}$ is the modulus of the roots and $\theta = \tan^{-1}(d/c)$ is the argument. The advantage of this form of the solution is that the constants of integration, and indeed, all the expressions involved in this definition, are now real numbers. This makes it easier to evaluate in practice. The equivalence of the two expressions for the solution in the case of complex roots is not obvious and requires a considerable amount of algebra to demonstrate. We therefore leave the proof that this is the case to an appendix.

Case 3: Repeated Roots

If $a_1^2 + 4a_2 = 0$, then the roots are real but not distinct, that is, we have $\lambda_1 = \lambda_2 = \lambda = a_1 / 2$. In this case the solution takes the form

$$y_n = (C_1 + C_2 n)\lambda^n.$$

We can easily demonstrate that this is a valid solution by substituting it back into the original equation.

EXAMPLE(S)

Find the general solutions for the following homogeneous difference equations

(a) $y_n = y_{n-1} - \frac{2}{9} y_{n-2}$

(b) $y_n = -2y_{n-1} - \frac{5}{4} y_{n-2}$

(c) $y_n = \frac{1}{2} y_{n-1} - \frac{1}{16} y_{n-2}$

For part (a), we have characteristic equation $\lambda^2 - \lambda + 2/9 = 0$ which gives us roots $\lambda_1 = 1/3$ and $\lambda_2 = 2/3$. Since the roots are real and distinct, the solution takes the form

$$y_n = C_1 \left(\frac{1}{3}\right)^n + C_2 \left(\frac{2}{3}\right)^n.$$

For part (b), we have characteristic equation $\lambda^2 + 2\lambda + 5/4 = 0$ which gives us roots $\lambda_{1,2} = 1 \pm i/2$. The roots are complex conjugates with modulus $\sqrt{1+(1/2)^2} = 1.118$ and argument $\tan^{-1}(1/2) = 0.4636$. The solution therefore takes the form

$$y_n = (1.118)^n \left(C_1 \cos(0.4636n) + C_2 \sin(0.4636n)\right).$$

For part (c), we have characteristic equation $\lambda^2 - (1/2)\lambda + (1/16) = 0$ which gives us roots $\lambda_1 = \lambda_2 = 1/4$. Since the roots are real but not distinct, the solution takes the form

$$y_n = (C_1 + C_2 n)\left(\frac{1}{4}\right)^n.$$

If the equation we wish to solve is nonhomogeneous, then the general solution is found by taking the sum of the general solution to the associated homogeneous problem and a particular integral. The method by which we find a particular integral will usually involve an initial guess as to the form of the equation followed by the use of the method of undetermined coefficients to find specific values for its parameters. For example, if the nonhomogeneous part of the equation is simply a constant then we have an equation of the form

$$y_n = a_1 y_{n-1} + a_2 y_{n-2} + a_0.$$

Since the nonhomogeneous part of the equation simply consists of the constant a_0, it is reasonable to assume a particular integral which is itself a constant. We therefore guess a solution of the form $y_p = c$ and look for a specific value of c using the method of undetermined coefficients. Substituting $y_p = c$ into the equation gives us

$$c = a_1 c + a_2 c + a_0 \Rightarrow c = \frac{a_0}{1 - a_1 - a_2}$$

If the roots of the characteristic equation are real and distinct, we can combine the general solution of the homogeneous equation and the particular integral we have just found to write down a general solution for the nonhomogeneous equation which takes the form

$$y_n = C_1 \lambda_1^n + C_2 \lambda_2^n + \frac{a_0}{1 - a_1 - a_2}.$$

EXAMPLE

Find the general solution of the nonhomogeneous difference equation $y_n = 0.75 y_{n-1} - 0.125 y_{n-2} + 100.$

The characteristic equation is $\lambda^2 - 0.75\lambda + 0.125 = 0$ which factorizes to give $(\lambda - 0.5)(\lambda - 0.25) = 0$. The roots are therefore 0.5 and 0.25 and the general solution of the homogeneous equation is $y_n = C_1 (0.5)^n + C_2 (0.25)^n$. Assuming a particular integral of the form $y_p = c$, we solve for the unknown parameter c to get $y_p = 100/(1 - 0.75 + 0.125)$ which gives us $y_p = 800/3$. The general solution of the nonhomogeneous equation is therefore

$$y_n = C_1 (0.5)^n + C_2 (0.25)^n + \frac{800}{3}.$$

In general, when solving for a particular integral, we assume a functional form which is similar to the nonhomogeneous part of the equation. For example, in the following case we have $f(n)$ equal to a linear function of n. Therefore, we assume a particular integral which takes the general form $y_p = a + bn$, where a and b are unknown parameters.

EXAMPLE

Find the particular solution of the equation $y_n = \frac{1}{4} y_{n-2} + 1 + 2n$ with initial conditions $y_0 = y_1 = 1$.

We have characteristic equation $\lambda^2 - 1/4 = 0$, and therefore the roots are $\lambda_1 = 1/2$ and $\lambda_2 = -1/2$, and the general solution of the homogeneous part of this equation takes the form $y_n = C_1(1/2)^n + C_2(-1/2)^n$. Assuming a particular integral of the form $y_p = a + bn$, we use the method of undetermined coefficients to write

$$a + bn - \frac{1}{4}(a + b(n-2)) = 1 + 2n.$$

$$\left(\frac{3}{4}a + \frac{3}{2}b\right) + \left(\frac{3}{4}b\right)n = 1 + 2n$$

Equating coefficients now gives us $a = -4/9$ and $b = 8/3$. The general solution of the nonhomogeneous equation therefore takes the form

$$y_n = C_1\left(\frac{1}{2}\right)^n + C_2\left(-\frac{1}{2}\right)^n - \frac{4}{9} + \frac{8}{3}n.$$

From our initial conditions we obtain a pair of simultaneous equations in C_1 and C_2 which take the form

$$C_1 + C_2 - \frac{4}{9} = 1$$
$$\frac{C_1}{2} - \frac{C_2}{2} - \frac{4}{9} + \frac{8}{3} = 1.$$

These can be solved to give us $C_1 = -1/2$ and $C_2 = 35/18$. Therefore, the particular solution which is consistent with these initial conditions is given by the equation

$$y_n = -\frac{1}{2}\left(\frac{1}{2}\right)^n + \frac{35}{18}\left(-\frac{1}{2}\right)^n - \frac{4}{9} + \frac{8}{3}n.$$

Next, let us consider an example of a second-order difference equation from Economics. The Samuelson multiplier-accelerator model of the business cycle provides a good example of the use of a second-order difference model in economic analysis. In this model, lags in adjustment of consumption and investment expenditures act to generate business cycles. The model is summarized in three key equations

$$Y_t = C_t + I_t + G_t$$

$$C_t = cY_{t-1}$$

$$I_t = v(C_t - C_{t-1}).$$

The first of these equations is the national income accounting identity. It states that total output is the sum of private section expenditure on consumption goods (C), investment goods (I), and government consumption (G). The second equation is the consumption function which states that private consumption expenditures are proportional to national output with a one period lag. The third equation is the investment function which states that investment adjusts according to the lagged change in private consumption expenditures.

Now, let us assume that the parameters c and v take the values $c = 0.8$ and $v = 1.25$. We will also assume that government spending is constant and equal to 100. This gives us a difference equation of the form

$$Y_t = 1.8Y_{t-1} - Y_{t-2} + 100. \tag{11.6}$$

The characteristic polynomial for this equation is $\lambda^2 - 1.8\lambda + 1 = 0$ which has roots $\lambda_{1,2} = 0.9 \pm 0.4359i$. The modulus is therefore equal to $\sqrt{0.9^2 + 0.43592^2} = 1$, and the argument is $\theta = \tan^{-1}(0.4359/0.9) = 0.451$. We can therefore write the complementary function as

$$Y_t^c = C_1 \cos(0.451t) + C_2 \sin(0.451t).$$

Note that the fact that the modulus is equal to one means that this particular configuration of the model will produce stable cycles. A particular solution can be found by solving equation (11.6) for a constant level of output. This gives us $Y_t^p = 100/0.2 = 500$, and this, in turn, allows us to write the general solution of the nonhomogeneous equation as

$$Y_t^g = C_1 \cos(0.451t) + C_2 \sin(0.451t) + 500.$$

We need a pair of boundary conditions to solve for the two constants in this expression. For example, let us assume that $Y_0 = Y_{-1} = 450$. This gives us a pair of equations of the form

$$450 = C_1 + 500$$
$$450 = C_1 \cos(-0.451) + C_2 \sin(-0.451)$$

which can be solved to give us $C_1 = -50$ and $C_2 = 11.47$. This means that we can write the particular solution of the model which is consistent with the initial conditions as

$$Y_t = -50 \cos(0.451t) + 11.47 \sin(0.451t) + 500.$$

Therefore, with the parameter values we have assumed, and these initial conditions, the model produces stable cycles around the equilibrium value $Y = 500$. This is illustrated in the plot of the time path of output shown in Figure 11.1

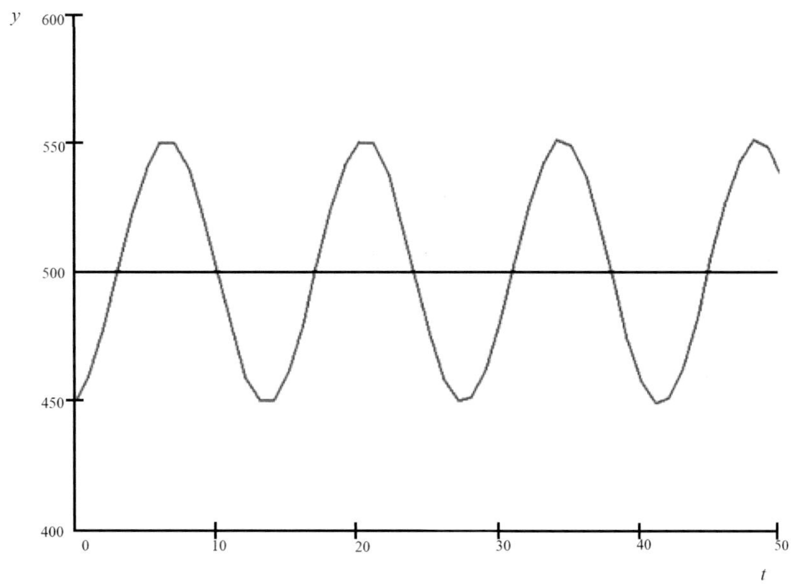

FIGURE 11.1 Time path of output for Samuelson multiplier-accelerator model with complex roots.

We should note that stable cycles are only produced for very particular combinations of the parameter values. Small changes in either the consumption or investment parameter will alter the nature of the solution so that either the cycles become damped or explosive. If the roots are complex, then we can show that, for general parameter values, the modulus is equal to $\sqrt{cv}$. It follows that, if the product of cv is greater than one, then the solution is explosive, while, if it is less than one, then the solution is damped. It is only in the special case that $cv = 1$ that the solution consists of a stable cycle. The proof of this is left as an exercise for the interested reader.

REVIEW EXERCISES – SECTION 11.2

1. Find the general solution for each of the following homogeneous equations

(a) $y_n = y_{n-1} + 2y_{n-2}$

(b) $y_n = -2y_{n-1} - 5y_{n-2}$

(c) $y_n = \frac{2}{3}y_{n-1} - \frac{1}{9}y_{n-2}$

2. Find the general solution for each of the following nonhomogeneous equations

(a) $y_n = -\frac{1}{6}y_{n-1} + \frac{1}{6}y_{n-2} + 2$

(b) $y_n = -y_{n-1} - \frac{5}{4}y_{n-2} + 3$

3. Find the particular solution of the following nonhomogeneous equation which is consistent with the initial conditions given

$$y_n = \frac{2}{5}y_{n-1} - \frac{1}{25}y_{n-2} + 5 \qquad y_0 = 0$$

$$y_1 = 1$$

4. In the case of complex roots, show that the nature of the general solution of $Y_t = c(1+v)Y_{t-1} - cvY_{t-2} + f(t)$ depends on the value of cv, where $cv = 1$ implies stable cycles, $cv > 1$ implies explosive cycles, and $cv < 1$ implies damped cycles.

11.3 SOLUTION BY BACKWARD SUBSTITUTION

Backward substitution provides an alternative method for the solution of first-order difference equations, which can be generalized to give a method for solving equations of any order.

Another method for solving difference equations that works well for first-order linear equations is that of *backward substitution*. Consider the general first-order nonhomogeneous equation defined in equation (11.2). We have $y_n = a_1 y_{n-1} + a_0$ and lagging each term in this expression will give us $y_{n-1} = a_1 y_{n-2} + a_0$ which we can substitute for y_{n-1} in the original expression. Moreover, we can continue to do this indefinitely, each time replacing a lagged term y_{n-k} with a term of the form y_{n-k-1}. This process is summarized below

$$\begin{aligned} y_n &= a_1 y_{n-1} + a_0 \\ &= a_1^2 y_{n-2} + a_1 a_0 + a_0 \\ &= a_1^3 y_{n-3} + a_1^2 a_0 + a_1 a_0 + a_0 \\ &\dots \\ &= a_1^n y_0 + a_0 \sum_{i=1}^{n} a_1^{i-1} = a_1^n y_0 + \frac{a_0\left(1 - a_1^n\right)}{1 - a_1}. \end{aligned}$$

This is the same as the particular solution for the equation that we derived in Section 11.1 for initial value of y equal to y_0.

EXAMPLE

Consider the following model drawn from economic theory. Output Y is equal to the sum of consumption expenditures C and investment I, which is assumed to be constant. Consumption depends on the level of output but with a one-period lag. We can therefore write down a simple model of output determination as

$$\begin{aligned} Y_t &= C_t + I_t \\ C_t &= cY_{t-1} \\ I_t &= \overline{I}. \end{aligned}$$

The parameter c is the *marginal propensity to consume* or MPC which we assume is greater than zero but less than one. Combining these equations allows us to write the model as a linear first-order difference equation.

$$Y_t = cY_{t-1} + \overline{I}.$$

Backward substitution allows us to write this equation as

$$Y_t = \overline{I}\left(1 + c + c^2 + \dots c^t\right) + c^t Y_0 = \overline{I}\left(\frac{1 - c^t}{1 - c}\right) + c^t Y_0.$$

Since we have assumed that $0 < c < 1$, it follows that, as $t \to \infty$, $Y_t \to \bar{I}/(1-c)$. The expression $1/(1-c)$ is a familiar expression from macroeconomic theory, where it is referred to as the *Keynesian expenditure multiplier*. This measures the effects of an increase in exogenous or autonomous expenditures on national output.

The method of backward substitution offers an interesting alternative to the solution method we set out in Section 11.1, but does it really add anything more? In terms of difficulty, the two methods are about the same and when it comes to higher-order difference equations, the method of backward substitution becomes considerably more unwieldy and expensive in terms of the extra algebra it requires. We can show, however, with the addition of a useful device drawn from matrix algebra, the method of backward substitution becomes a very efficient way of solving higher-order equations.

Let us consider the general second-order linear equation with constant coefficients. That is, an equation of the form $y_n = a_1 y_{n-1} + a_2 y_{n-2} + a_0$. An alternative way to present this equation is as a first-order matrix equation, as shown in equation (11.7)

$$\begin{bmatrix} y_n \\ y_{n-1} \end{bmatrix} = \begin{bmatrix} a_1 & a_2 \\ 1 & 0 \end{bmatrix} \begin{bmatrix} y_{n-1} \\ y_{n-2} \end{bmatrix} + \begin{bmatrix} a_0 \\ 0 \end{bmatrix}. \tag{11.7}$$

Moreover, this representation generalizes further. Consider a general difference equation of order m. We can write this as a single equation in the form

$$y_n = \sum_{i=1}^{m} a_i y_{n-i} + a_0$$

or in matrix form as $z_n = Az_{n-1} + w$, where the vectors z and w, and the matrix A are defined as follows

$$A = \begin{bmatrix} a_1 & a_2 & \cdots & \cdots & a_m \\ 1 & 0 & \cdots & \cdots & 0 \\ 0 & 1 & \cdots & \cdots & 0 \\ \vdots & \vdots & \ddots & \ddots & \vdots \\ 0 & 0 & \cdots & 1 & 0 \end{bmatrix},\ z = \begin{bmatrix} y_n \\ y_{n-1} \\ \vdots \\ \vdots \\ y_{n-m} \end{bmatrix},\ w = \begin{bmatrix} a_0 \\ 0 \\ \vdots \\ \vdots \\ 0 \end{bmatrix}.$$

When we write our equation in matrix form like this, it becomes straightforward to solve equations of *any* order using the method of backward substitution. The solution takes the form

$$z_n = A^n z_0 + \sum_{i=0}^{n-1} A^i w .$$

EXAMPLE

To solve the difference equation $y_n = \frac{1}{12} y_{n-1} + \frac{1}{12} y_{n-2} + 2$ with initial conditions $y_0 = 3$ and $y_{-1} = 4$, we first write it as a first-order matrix equation.

$$z_n = Az_{n-1} + w \qquad z_n = \begin{bmatrix} y_n \\ y_{n-1} \end{bmatrix}, A = \begin{bmatrix} 1/12 & 1/12 \\ 1 & 0 \end{bmatrix}, w = \begin{bmatrix} 2 \\ 0 \end{bmatrix}$$

The solution takes the form $z_n = A^n z_0 + \sum_{i=0}^{n-1} A^i w$, where $z_0 = \begin{bmatrix} 4 & 3 \end{bmatrix}^T$. Using this expression, we can easily calculate the value of y for any value of n. For example, we have

$$z_2 = \begin{bmatrix} 2.5486 \\ 2.5833 \end{bmatrix},$$

which gives us both $y_2 = 2.5486$ and $y_1 = 2.5833$.

REVIEW EXERCISES – SECTION 11.3

1. Solve the difference equation $y_t = 0.2 y_{t-1} + 0.8$ by the method of backward substitution and show that $y_t \to 1$ as $t \to \infty$.
2. For the general difference equation $y_t = a y_{t-1} + b$, show that $y_t \to b/(1-a)$ as $t \to \infty$ if $-1 < a < 1$ but is unstable otherwise.

11.4 BOUNDARY CONDITIONS AND EXPECTATIONS

Boundary conditions are not limited to initial values. In this section, we show how a simple model of asset prices generates a boundary condition which depends on a terminal value rather than an initial value for the variable of interest.

The general solution of a difference equation always contains arbitrary constants of integration, the number of which depends on the order of the equation, with a first-order equation containing a single constant of integration, a second-order equation containing two, and so on. To eliminate these, we rely on some form of boundary condition. For many problems, the boundary conditions consist of starting values or initial conditions. Indeed, the terms boundary condition and initial condition are often used almost synonymously. However, there is an important class of problem in economics for which this is not the case. These are models in which the current value of a variable of interest depends on its expected future value. In such cases, the boundary condition often depends on the future value of the variable of interest rather than its initial value.

Let us consider the example of the price of a financial asset such as a company share. The theory of asset pricing states the market will be in equilibrium when the (risk adjusted) return on holding an asset is equal to the return on the market as a whole. We can write this condition as follows

$$\frac{d + p_{t+1}^{e} - p_t}{p_t} = r \tag{11.8}$$

where p is the asset price, d is the dividend, and r is the market return. For simplicity, we assume that the dividend and the market return are constants. The one-period return on holding the asset depends on the dividend and the expected change in the price during the holding period.

Assuming perfect foresight, so that $p_{t+1}^{e} = p_{t+1}$, we can solve (11.8) to obtain a first-order difference equation of the form

$$p_{t+1} = (1+r)p_t + d\,.$$

This has general solution

$$p_t = C(1+r)^t + \frac{d}{r} \tag{11.9}$$

where C is a constant of integration. The term d/r in equation (11.9) reflects the market fundamentals of the asset in question. It is equal to the discounted present value of the stream of dividends with a discount rate given by the market rate of return. Clearly, if the asset price is to be determined by market fundamentals, then we require $C = 0$. We therefore have a rather uninteresting solution to our difference equation in which the asset price is simply equal to the market fundamental rate, and there is no dynamic adjustment of any kind.

The simple solution described in the previous paragraph applies only if the dividend level and the market rate of return are either constant, or change suddenly and without warning, so that the asset price adjusts immediately. In cases where a change in d and/or r is anticipated at some stage in the future, then we get a rather more interesting solution. Let us consider a case in which r is constant but, at date t_1, the market becomes aware that, at a future date t_2, the dividend rate is likely to rise from d_1 to d_2. Up to date t_1 the price of the share is determined by its market fundamental rate $p_1 = d_1 / r$, and after t_2 it will be determined by the new market fundamental rate $p_2 = d_2 / r$. The interesting question however, is what happens between these dates, that is once the market becomes aware of the future change, but before that change actually takes place.

Let us consider two possible responses to the change in market fundamentals and show that neither of these is likely to happen in practice. First, if there is no change in price at date t_1, then market traders will lose out on a profitable opportunity. The fact that dividends are going to rise in the future means that the price of the asset will rise, and there is therefore an opportunity to make a profit by purchasing it immediately. No change in price is therefore inconsistent with the assumption that market traders will look to exploit any profit opportunities available to them. If a constant price is not consistent with profit maximization, then will the price of the asset jump immediately to its new equilibrium value? Again, this is not consistent with profit maximizing behavior. During the interim period t_1 to t_2, dividends are lower than those on other assets. Traders could therefore make a higher return by holding these alternative assets.

If no change, and immediate change to the new equilibrium are both ruled out, how can we determine the value of the asset during the period between the market becoming aware of the change, and the change actually taking place. To do this, let us go back to the general solution of the difference equation (11.9). We know that after t_2 the equilibrium price is equal to $p_2 = d_2 / r$. We can therefore use this as a boundary condition to solve for the constant of integration. we have

$$\frac{d_2}{r} = C(1+r)^{t_2} + \frac{d_1}{r} \Rightarrow C = \left(\frac{d_2 - d_1}{r}\right)(1+r)^{-t_2}.$$

We are now able to set out a complete solution for the price of the asset. Given the assumptions we have made, we have

$$p_t = \begin{cases} d_1 / r & t < t_1 \\ (1+r)^{t-t_2}\left(\dfrac{d_2 - d_1}{r}\right) + \dfrac{d_1}{r} & t_1 \leq t < t_2 \\ d_2 / r & t \geq t_2 \end{cases}$$

This defines the complete time path for the price of the asset from the period $t < t_1$ before agents become aware that a change in dividends will take place, followed by the period $t_1 \leq t < t_2$ during which agents are aware that a change will happen but before it actually takes place, and finally, the period $t \geq t_2$ when the change has actually occurred. Note that, in solving for the constant of integration, we have used a boundary condition which depends on the future value of the variable of interest rather than an initial condition. The boundary condition here requires that the solution path be such that the price of the asset reach its new equilibrium value on the date at which the change in dividend actually takes place. A jump in the asset price at that date is not consistent because it would imply market traders ignoring a profitable opportunity.

EXAMPLE

Let $d_1 = \$100$ and $d_2 = \$120$ and let the market rate of return $r = 0.05$. At date $t_1 = 10$ information becomes available that the dividend rate will rise from d_1 to d_2 at $t_2 = 30$. The equilibrium price of the asset will rise from $p_1 = \$100 / 0.05 = \$2,000$ for $t < 10$ to $p_2 = \$120 / 0.05 = \$2,400$ for $t \geq 30$. Between these dates the price of asset adjusts according to the equation

$$\begin{aligned} p_t &= (1+r)^{t-t_2}\left(\frac{d_2 - d_1}{r}\right) + \frac{d_1}{r} \\ &= (1.05)^{t-30} \times \$400 + \$2,000 \end{aligned}$$

The time path of the equity price is illustrated in Figure 11.2. This shows that there is an initial jump in the price when new information about the future dividend rate becomes available but there is no jump when the actual change in the dividend rate takes place.

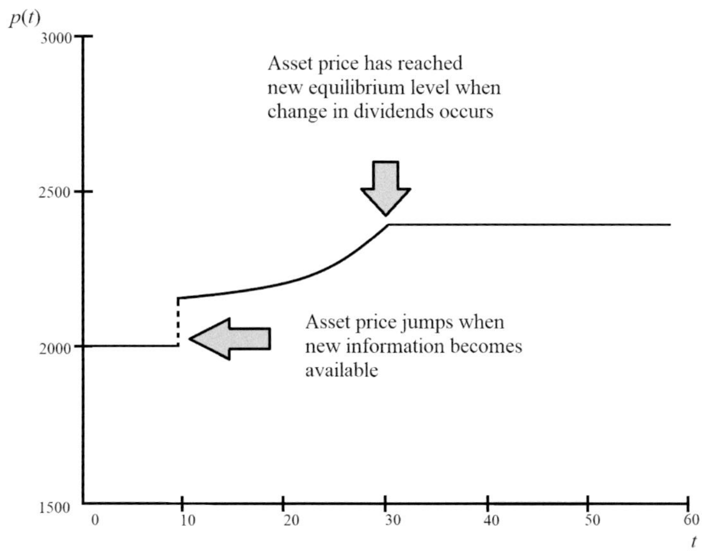

FIGURE 11.2 Time path of asset price in response to change in expectations.

REVIEW EXERCISES – SECTION 11.4

1. Consider an asset that bears constant dividend $d = \$10$. The market rate of return r is equal to 5%. At $t = 0$ information becomes available that the dividend will increase to \$15. Calculate the size of the immediate jump in the price of the asset when the date of the increase is as follows

(a) $t = 1$

(b) $t = 2$

(c) $t = 10$.

2. An asset bears a dividend of \$10. Determine the price of the asset over the period $t = 0$ to $t = 10$ if the market rate of return is initially equal to 10% but, at date $t = 2$ agents become aware that it will fall to 5% at $t = 5$.

APPENDIX: SOLUTION FOR THE CASE OF COMPLEX ROOTS

In the main text, we state that the solution for the homogeneous second-order difference equation with complex roots $\lambda_1 = a + bi$ and $\lambda_2 = a - bi$ can be written in the form

$$y_n = r^n \left(C_1 \cos(\theta n) + C_2 \sin(\theta n)\right)$$

where $r = \sqrt{a^2 + b^2}$ and $\theta = \tan^{-1}(b/a)$. This is more convenient than the form

$$y_n = A_1 \lambda_1^n + A_2 \lambda_2^n$$

in which the both the roots λ_1 and λ_2, and the weights A_1 and A_2, are complex conjugates since it includes only real expressions and can therefore be more easily evaluated. The proof of this result is now given as follows.

By De Moivre's theorem, we can write

$$\lambda_1 = r\cos(\theta + i\sin\theta)$$
$$\lambda_2 = r\cos(\theta - i\sin\theta).$$

Therefore, we can write

$$y_n = A_1 r^n \left(\cos(\theta n) + i\sin(\theta n)\right) + A_2 r^n \left(\cos(\theta n) - i\sin(\theta n)\right)$$
$$= (A_1 + A_2) r^n \cos(\theta n) + i(A_1 - A_2) r^n \sin(\theta n).$$

A_1 and A_2 are complex conjugates, so let $A_1 = c + di$ and $A_2 = c - di$. This means that we can eliminate all the complex terms from our solution and write it in the form

$$y_n = r^n \left(C_1 \cos(\theta n) + C_2 \sin(\theta n)\right)$$

where $C_1 = 2c$ and $C_2 = -2d$. This expression makes it clear that the stability of the solution depends on the condition that the modulus r must be less than one in absolute value.

EXAMPLE

The second-order difference equation $y_n = \frac{1}{2} y_{n-1} - \frac{5}{16} y_{n-2}$ has roots $\lambda_1 = \frac{1}{4} + \frac{1}{2} i$ and $\lambda_2 = \frac{1}{4} - \frac{1}{2} i$. These roots have modulus $r = \sqrt{\left(\frac{1}{4}\right)^2 + \left(\frac{1}{2}\right)^2}$ and argument $\theta = \tan^{-1}(2)$. We can write the general solution of this equation in either of the two equivalent forms

$$y_n = A_1\left(\frac{1}{4}+\frac{1}{2}i\right)^n + A_2\left(\frac{1}{4}-\frac{1}{2}i\right)^n$$
$$y_n = 0.559^n\left(C_1\cos(1.1071n)+C_2\sin(1.1071n)\right)$$

Given initial conditions $y_0 = 0$ and $y_1 = 1$, we can solve these to get particular solutions in which the constants of integration are $A_1 = -i$, $A_2 = i$, and $C_1 = 0$, $C_2 = 2$. We can therefore write the particular solution as

$$y_n = -i\left(\frac{1}{4}+\frac{1}{2}i\right)^n + i\left(\frac{1}{4}-\frac{1}{2}i\right)^n$$
$$y_n = 2\left(0.559^n\right)\sin(1.1071n).$$

The second form is generally preferable for computational purposes because it does not require any calculations with complex numbers.

APPENDIX A

CODING IN PYTHON

VARIABLE TYPES

Strings

The most common types of variables you will come across when coding in Python are *strings*, *integers*, and *floats*.

A *string* is a variable that consists of a block of text. You can then print them using the **print**() command. This is the basic command used to output results to the screen. We can use this command for all the different types of variables defined in Python.

For example

```
a = "This is a Python string"
print(a)
```

will produce the following output

```
This is a Python string
```

We can add strings together to create new strings. For example, if we run the following block of code

```
a = "My name is "
b = "John Smith"
c = a+b
print(c)
```

then we will get the following

```
My name is John Smith
```

Note that Python will accept single or double quotation marks so the following definitions are equally valid

```
a = "This is a Python string"
a = 'This is a Python string'
```

Integers

Integers are whole numbers which can be positive, negative, or zero. We can assign values to variable names using the equals sign. For example, $a=1$, simply defines an integer variable a and assigns to value 1 to it. We can perform standard arithmetic operations such as addition and subtraction on integer variables to create new integers. For example

```
a = 1
b = 2
c = a+b
print("The sum of ", a, "and ", b,"is ",c)
```

produces the following output.

```
The sum of  1 and  2 is  3
```

Floating-point Numbers

Floating-point numbers, or *floats*, are variables that can be represented in decimal form. In Python, they provide a way of representing real numbers. We can define them in the standard way by using the equals sign. For example, $a = 1.5$, assigns the value 1.5 to the variable a. We can perform all the standard arithmetic operations on floating-point numbers. For example, the following code divides an integer a by another integer b, with the outcome being a floating-point number c.

```
a = 1
b = 2
c = a/b
print("If we divide ", a, "by ", b,"then we get ",c)
```

This produces the following output

```
If we divide  1 by  2 then we get  0.5
```

Converting Variable Types

It is sometimes necessary to convert a variable of one type into another. We can convert either integers or floating-point numbers to strings using the **str**() function. This can be useful when combining (or 'concatenating') strings to output results. For example, the following code takes a floating-point number *pi,* and converts it to a string, so that we can output a result.

```
pi = 3.142
a ="The value of pi is approximately "
b = str(pi)
c = a+b
print(c)
```

This gives the following output

```
The value of pi is approximately 3.142
```

Similarly, if *a* is a string which represents a number, then we can convert it to either an integer or a floating-point number using the int() or float() commands. The following code illustrates this process

```
a = "10"
b = "10.15"
a_int = int(a)
b_float = float(b)
print("a_int = ", a_int)
print("b_float = ",b_float)
```

Executing this code gives us the following

```
a_int =  10
b_float =  10.15
```

INPUT AND OUTPUT

The main input and output commands in Python are *input* and *print*. These work as follows: *input* invites the user to enter some information, which can be a number or text, while *print* sends output to the screen.

EXAMPLE

The following code asks the user to input a number. The program then takes the square of this number and returns it to the screen, along with an explanation of what it has done.

```
x=input("Input a number ")
x1=float(x)
z=x1**2
print("The squared value of your number is ",z)
```

If we run this code, then we obtain the following output.

```
Input a number 5
The squared value of your number is  25.0
```

Note that the default is for Python to treat interactive input of this kind as a string. Before we can perform any numerical operations on our input, we must convert it to a number. Here we have used the **float**() command to convert our input to a floating-point decimal number. A less general alternative is the **int**() command, which can be used if the input number is an integer.

FORMATTING OUTPUT

When working with floating-point numbers, we often wish to limit the number of decimal places in our output. For example, an irrational number such as 1/6 has an infinite decimal representation. Obviously, Python cannot report an infinite number of digits but it will typically report more than we wish. To limit the number of digits, we use the following command

```
print("{:.4f}".format(1/6))
```

The expression "{:.4f}" is a formatting statement that indicates that we would like the results to be reported to an accuracy of four decimal places. To compare what happens when we use this command and when we leave the output unformatted, see what happens when we run the following code.

```
a = 1/6
print(a)
print("{:.4f}".format(a))
```

The output from this code is

```
0.16666666666666666
0.1667
```

The unformatted print command reports the number 1/6 to seventeen decimal places. The formatted command reports it to four decimal places and rounds the last digit appropriately. In general, it is good practice to format output to make interpretation of the results easier for the user.

CONDITIONAL STATEMENTS

Conditional statements instruct the computer to alter how the code is executed, depending on the truth, or otherwise of a statement. They always begin with a statement of the form *if* <something is true> following by a colon (:). The code which follows instructs the computer to execute statements based on the truth of the *if* statement. It is also possible to modify the code further by the use of *elif* statements, which allow for further conditions to be assessed.

EXAMPLE

The following code asks the user to input a number. The program then returns a statement as to whether this number is greater than, equal to, or less than the number five.

```
x=input("Input a number ")
x1=float(x)
if x1>5:
    print("Your number is greater than five")
elif x1==5:
    print("Your number is equal to five")
elif x1<5:
    print("Your number is less than five")
```

Running this code gives the following output.

```
Input a number 6
Your number is greater than five
```

FOR LOOPS

For loops instruct the computer to execute a block of code a fixed number of times. These loops have the following general structure.

for idx in range (a,b):

<execute some code>

Starting with the value a, the default is to increase the *idx* by one unit until it reaches the value b-1. However, this can be modified to change the increment to different values if desired.

Note that we need to be careful in specifying the end-point of the range. For example, if we wish to perform a set of calculations for idx = 1,2,3,4,5, then we need to specify the end of the range as 6.

EXAMPLE

The following code calculates the cubed value of the integers 1,2,3,4, and 5, and prints the results to the screen.

```
print("The value of x cubed for x = 1 to 5 is ")
print("x","x-cubed")
for idx in range(1,6):
    print(idx,idx**3)
```

The output for this code is as follows.

```
The value of x cubed for x = 1 to 5 is
x x-cubed
1 1
2 8
3 27
4 64
5 125
```

WHILE LOOPS

For loops perform a fixed number of iterations of the code contained within the statements. Sometimes, however, we do not know in advance how many iterations will be needed to achieve a given objective. The *while* loop structure instructs the program to continue looping until a desired objective is achieved. The general structure of such loops is as follows.

<initial condition>

while <condition is true>:

<execute some code>

<modify condition>

EXAMPLE

The following code finds an approximate value for the square root of the number five.

```
x=2
z=x**2-5
while z<0:
    print("{:.4f}".format(x),"{:.4f}".format(z))
    x=x+0.1
    z=x**2-5
print("{:.4f}".format(x),"{:.4f}".format(z))
```

If we run this code, then we get the following output.

```
2.0000 -1.0000
2.1000 -0.5900
2.2000 -0.1600
2.3000 0.2900
```

This tells us that the square root of five lies somewhere between 2.2 and 2.3 because the value of the expression $z = x^2 - 5$, changes sign between these two values of x. This took three iterations through the loop to find this result.

Note that we have used a formatting command to make the computer print the output to four decimal places. This command takes the form

```
print("{:.4f}".format(x),"{:.4f}".format(z))
```

Without the formatting statement, the default output would consist of the full decimal expression of the number which may consist of a very long string of numbers following the decimal point. It is usually good practice to control the way in which numbers are presented to avoid this happening and to make the output easy to read and interpret.

Note that it is very easy to get stuck in infinite loops when using this particular structure because we have many situations in which the condition will never be met. For example, the following loop will theoretically go on forever, since the condition $x^2 > 0$, will always be satisfied.

```
x = 1
z= x**2
while z>0:
    print(x,z)
    x = x+1
    z = x**2
```

It is therefore advisable to put in some sort of control to exit from the loop if it is taking too many iterations to meet the condition. This can be done using an *if* statement which is conditional on the *counter* variable used to count how many times the code has gone through the loop.

```
counter = 0
x = 1
z= x**2
while z>0:
    counter = counter+1
    print(x,z)
    x = x+1
    z = x**2
    if counter>10:
        print("Loop terminated after ",counter," iterations")
        break
```

This produces the following output.

```
1 1
2 4
3 9
4 16
5 25
6 36
7 49
8 64
9 81
10 100
11 121
Loop terminated after  11  iterations
```

APPENDIX B

ODD NUMBERED EXERCISES ANSWERS

SECTION 1.1

1. **(a)** 0.25 can be written as ¼. It is, therefore, a rational number and belongs to the sets $\mathbb{Q}$ and $\mathbb{R}$. However, it does not belong to the sets $\mathbb{N}$ or $\mathbb{Z}$.

   (b) $\sqrt{2}$ is an irrational number. Therefore, $2\sqrt{2}$ belongs in the set $\mathbb{R}$ but not in $\mathbb{N}$, $\mathbb{Z}$ or $\mathbb{Q}$.

   (c) -4 is a negative integer. Therefore, it belongs in the sets $\mathbb{Z}$, $\mathbb{Q}$ and $\mathbb{R}$ but not in $\mathbb{N}$.

   (d) 0.666... is the decimal representation of the fraction 2/3. Hence, it belongs in the sets $\mathbb{Q}$ and $\mathbb{R}$ but not in $\mathbb{N}$ or $\mathbb{Z}$.

   (e) 5,489,127 is a (very large!) positive integer. Therefore, it belongs in all the sets considered, that is, $\mathbb{N}$, $\mathbb{Z}$, $\mathbb{Q}$, and $\mathbb{R}$.

3. We have $\sqrt{8} = \sqrt{4\times 2} = 2\sqrt{2}$, since root $\sqrt{2}$ has been demonstrated to be irrational, it follows that $\sqrt{8}$ is irrational.

SECTION 1.2

1. (a) $4-\dfrac{(3-2)}{3}=4-\dfrac{1}{3}=\dfrac{11}{3}$

   (b) $2(3-4)=2\times-1=-2$

   (c) $\dfrac{2}{3}-\dfrac{(3+1)}{4}=\dfrac{2}{3}-\dfrac{4}{4}=-\dfrac{1}{3}$

   (d) $5\times4-\dfrac{3}{2}=20-\dfrac{3}{2}=\dfrac{37}{2}$

   (e) $6\div3(1+2)=6\div6=1$

SECTION 1.3

1. $x^2-4=0$ can be written $x^2=4$ since the right-hand side is positive; this will have real roots $x=\pm2$. For $x^2+4=0$, we have $x^2=-4$, since the right-hand side is negative, there are no real solutions, and we have $x=\pm2\sqrt{-1}=\pm2i$.

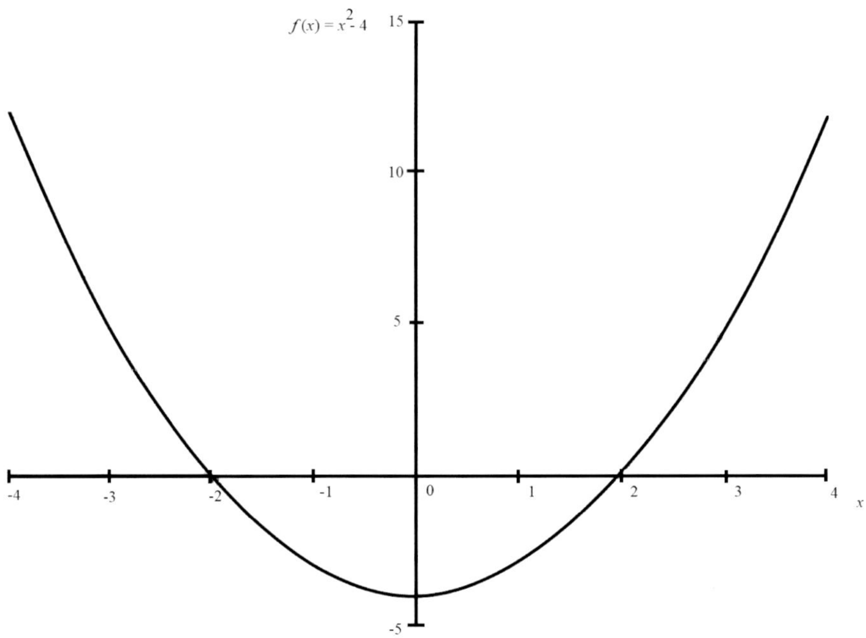

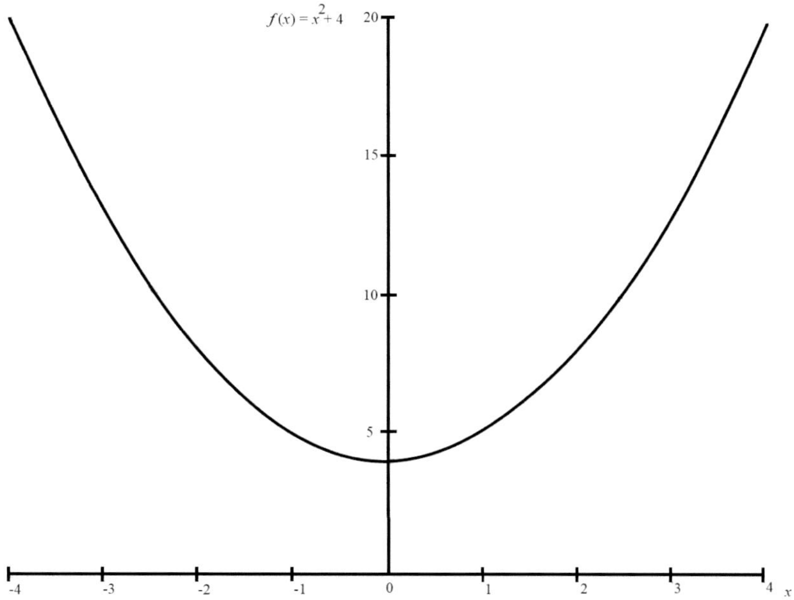

From the graphs, we note that when the roots are real and distinct, the graph cuts the horizontal axis in two places. If the roots are complex, then the function does not cut the horizontal axis at all.

3. Let $x = a + bi$ and $y = c + di$, we wish to find $z = e + fi$ such that $z = (x / y)$ or, alternatively, such that $yz = x$. Thus, we require

$$(c + di)(e + fi) = a + bi$$
$$(ec - df) + (cf + de)i = a + bi$$

This gives us a pair of simultaneous equations in e and f

$$ec - df = a \qquad (1)$$
$$cf + de = b \qquad (2)$$

These can be solved straightforwardly to yield

$$e = \frac{ac + bd}{c^2 + d^2} \qquad f = \frac{bc - ad}{c^2 + d^2}$$

which demonstrates the general result we require.

SECTION 1.4

1. **(a)** The set of real numbers greater than or equal to zero.
   (b) The set of real numbers less than zero.
   (c) All real numbers between −1 and +1 but not including these values.
   (d) The set of integer values −1, 0, 1, and 2.
   (e) The set of positive real numbers.

SECTION 1.5

1. **(a)** $(x+1)(x+2)=x^2+3x+2$
   (b) $(2x+1)(x+3)=2x^2+7x+3$
   (c) $(x+1)(x-1)=x^2-1$
   (d) $(x+3)^2=x^2+6x+9$
   (e) $x+x(x-1)=x^2$

SECTION 1.6

1. First, we modify the definition of the expression as shown below

```
def f(x):
  f = x**3-3*x
  return f
```

We then modify the limits as follows

```
xlower = -4
xupper = -1
```

Running the code now gives us the following result

```
There is a root at x = -1.73205081
```

SECTION 2.1

1. In each case calculate the slope of the function and then calculate the intercept using either pair of coordinates. If b is the slope and a is the intercept, we have

(a) $b=\frac{-5-(-1)}{3-1}=-2 \qquad a=-1-b\times 1=1 \qquad \Rightarrow y=1-2x$

(b) $b=\frac{11-7}{2-1}=4 \qquad a=7-b\times 1=3 \Rightarrow y=3+4x$

(c) $b=\frac{11-2}{4-1}=3 \qquad a=2-b\times 1=-1 \qquad \Rightarrow y=-1+3x$

3. If $x=4t-1 \Rightarrow t=\frac{x+1}{4}$. Substituting into the equation for y gives $y=3\left(\frac{x+1}{4}\right)$ or $y=\frac{3}{4}+\frac{3}{4}x.$

SECTION 2.2

1. (a) $y=\sqrt[3]{x}$ The domain is $-\infty<x<\infty$. The range is $-\infty<y<\infty$.

(b) $y=1/x^2$ The domain is $-\infty<x<\infty, x\neq 0$. The range is $-\infty<y<\infty, y\neq 0$.

(c) $y=|x|$ The domain is $-\infty<x<\infty$. The range is $0\leq y<\infty$.

(d) $y=-3x^2$ The domain is $-\infty<x<\infty$. The range is $-\infty<y\leq 0$.

3. (a) $|y|=x$ is not a functional relationship because some values of x are consistent with more than one value of y. For example, $x=1$ is consistent with $y=1$ and $y=-1$.

(b) $y=|x|$ is a functional relationship because every value of x is consistent with a unique value of y.

(c) $y^2-2x=0$ is not a functional relationship because some values of x are consistent with more than one value of y. For example, $x=2$ is consistent with $y=2$ and $y=-2$.

(d) $3y + 2x = 1$ can be written as a linear equation $y = \frac{1}{3} - \frac{2}{3}x$. This is a functional relationship because every real value of x produces a unique value of y.

SECTION 2.3

1. (a) $\lim\limits_{x\to\infty} \frac{1}{\sqrt{x}} = \frac{1}{\lim\limits_{x\to\infty}\sqrt{x}} = \frac{1}{\infty} = 0$

(b) $\lim\limits_{x\to 2}\left(x^2 + \frac{1}{x}\right)^3 = \left(\lim\limits_{x\to 2} x^2 + \lim\limits_{x\to 2}\frac{1}{x}\right)^3 = \left(4 + \frac{1}{2}\right)^3 = \frac{729}{2}$

(c) $\lim\limits_{x\to 0}\left(4x^2 + \frac{1}{x}\right) = \lim\limits_{x\to 0} 4x^2 + \lim\limits_{x\to 0}\frac{1}{x} = 0 + \infty = \infty$

3. We have $\lim\limits_{x\to 0}\frac{(2+x)^2 - 4}{x} = \lim\limits_{x\to 0}\frac{x^2 + 4x + 4 - 4}{x} = \lim\limits_{x\to 0}(4x + 4) = 4$

SECTION 2.4

1. (a) $f(x) = x^2 x^3 = x^5$

(b) $f(x) = \frac{x^2}{\sqrt{x}} = x^2 x^{-1/2} = x^{3/2} = \left(\sqrt{x}\right)^3$

(c) $f(x) = \left(4x^2\right)^{-2} = \frac{1}{\left(4x^2\right)^2} = \frac{1}{16x^4}$

(d) $f(x) = \sqrt{4x^{-2}} = \left(4x^{-2}\right)^{1/2} = \left(\frac{4}{x^2}\right)^{1/2} = \frac{2}{x}$

SECTION 2.5

1. (a) $f(2) = 4$

(b) $f(1) = 2$

(c) $f(0) = 1$

(d) $f(-1) = 1/2$

(e) $f(-2) = 1/4$

Sketching the function gives

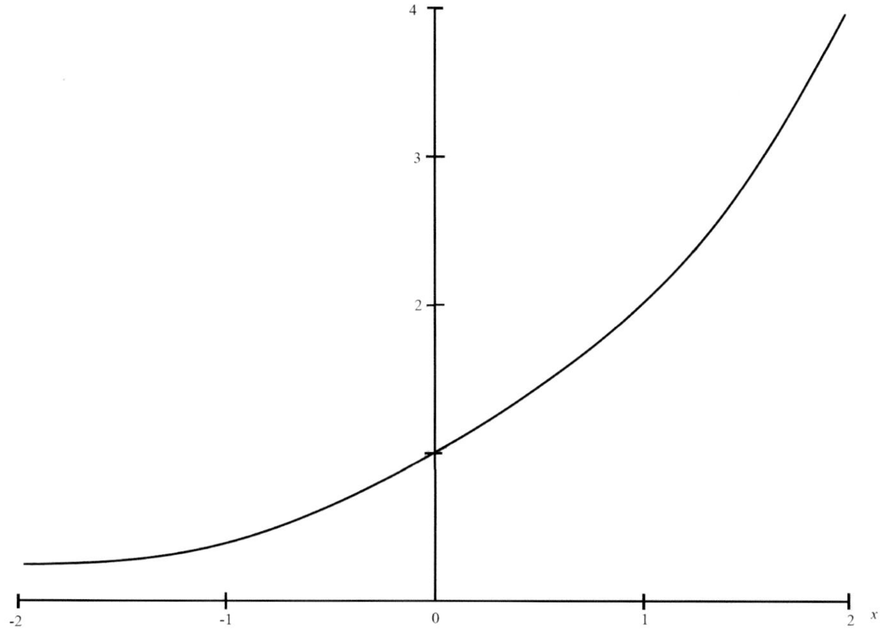

3. Note that $32 = 2^5$, therefore $\ln(32) = \ln(2^5) = 5\ln(2)$.

SECTION 2.6

1. (a) $f(x) = x^2 - 5x + 6 = (x-3)(x-2)$. Therefore, the roots are 3 and 2.

(b) $f(x) = x^2 - 6x + 9 = (x-3)^2$. In this case, there is a repeated root $x = 3$.

(c) $f(x) = 2x^2 + 3x + 2 = (2x+1)(x+1)$. Therefore, the roots are -1/2 and -1.

(d) $f(x) = 3x^2 + x - 2$. Solving the roots using the standard formula gives $x = 1/6$ and $x = -2/3$.

(e) $f(x) = x^2 - 4x + 5$. Solving for the roots using the standard formula gives $x = 2 \pm i$, that is, a pair of complex conjugates.

SECTION 2.7

1. To answer these questions, we first need to determine the length of the hypotenuse. Since the opposite and adjacent sides both have length 1 it follows that the hypotenuse has length $\sqrt{1^2+1^2}=\sqrt{2}$.

   (a) $x = 45^o = \dfrac{\pi}{4}$ radians

   (b) $\tan x = 1$

   (c) $\sin x = \dfrac{1}{\sqrt{2}}$

   (d) $\cos x = \dfrac{1}{\sqrt{2}}$

SECTION 3.1

1. For each of these questions, we calculate the slope as $\Delta y / \Delta x$ and then use either equation to calculate the intercept. This gives

   (a) $y = \dfrac{7}{5} + \dfrac{2}{5}x$

   (b) $y = 9 - 2x$

   (c) $y = 1 + 3x$

   (d) $y = 5$

3. **(a)** $x = -\dfrac{5}{3} + \dfrac{1}{3}y$

   (b) $x = -\dfrac{3}{2} - \dfrac{1}{2}y$

   (c) $x = \dfrac{5}{2} - \dfrac{1}{4}y$

   (d) $x = \dfrac{1}{2} - \dfrac{3}{4}y$

SECTION 3.2

1. To sketch these lines, we first solve for the equations in explicit form and then sketch the lines obtained. This gives the following.

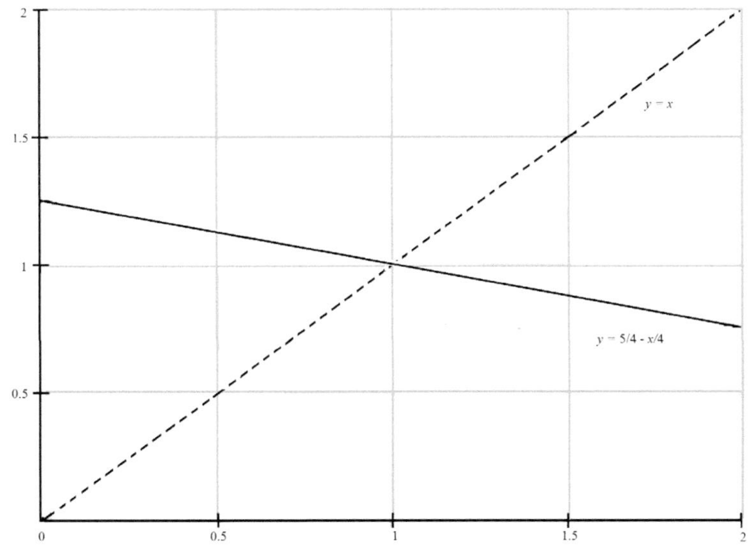

The lines cross approximately at the point $x = 1, y = 1$. We can confirm that this is the solution by substituting back into the original equations.

3. In each case, we establish that a unique solution exists by calculating the gradients for the lines and showing that they are not equal. A full solution is provided for the first question, with answers for the others.

(a) The slopes of the two equations are -3 and 1/2, respectively. Hence a unique solution exists.

To find the solution, we write the second equation in explicit form. This gives $x = -3 + 2y$. Substituting into the first equation gives

$$3(-3+2y)+y=5$$
$$-9+6y+y=5$$
$$7y=14$$
$$y=2$$

Substituting into the first equation gives $3x+2=5 \Rightarrow x=1$ which gives us the solution $x=1, y=2$.

(b) $x=1, y=2$

(c) $x=4, y=3$

(d) $x=2, y=5$

SECTION 3.3

1. **(a)** We have $p=102-2q$ and $q=48+p$. Substituting the first equation into the second equation gives $p=102-2(48+p)=6-2p$, and this gives $p=2$ as a solution. We can then solve for q using either equation; for example, using the supply curve gives us $q=48+2=50$.

   (b) $p=4,\ q=20$

   (c) $p=7,\ q=30$

3. We can write the system as

$$Y-C+M=350$$
$$-0.7Y+C=30$$
$$-0.4Y+M=10$$

We can now apply linear operations to write the system in triangular form. First, multiply equation 1 by 0.7 and add to equation 2.

$$Y-C+M=350$$
$$0.3C+0.7M=275$$
$$-0.4Y+M=10$$

Now, multiply equation 2 by 4/3 and add to equation 3 to obtain

$$Y-C+M=350$$
$$0.3C+0.7M=275$$
$$(7/3)M=1{,}550/3$$

This system is now in triangular form and we can solve for the endogenous variables. We have

$$M = \frac{1,550}{7} = 221$$
$$C = \frac{275 - 0.7 \times M}{0.3} = 400$$
$$Y = 350 + C - M = 528$$

where each number has been rounded to the nearest whole number.

SECTION 3.4

1. In each case, we use the method of substitution to eliminate y and then solve the polynomial function in x.

(a) $y = x^2 - 4x + 6,\ y = x$ gives $x^2 - 5x + 6 = 0$ which factorizes to give $(x-3)(x-2) = 0$. Therefore, the solutions are $x = 3, y = 3$ and $x = 2, y = 2$.

(b) $y = x^2 - 4x + 8, y = 4x - 8$ gives $x^2 - 8x + 16$ which factorizes to give $(x-4)^2 = 0$. Therefore, there is a repeated root and only one solution with $x = 4, y = 8$.

(c) $y = x^3 - x^2 + x - 2, y = 3x^2 - 4x$ gives $x^3 - 2x^2 + 3x - 2 = 0$. $x = 1$ is an obvious solution. Therefore, we extract this root and look to solve $(x-1)(x^2 - 3x + 2) = 0$. The quadratic expression factorizes to give $(x-1)(x-2)$. Therefore, one root is repeated and we have two solutions $x = 1, y = -1$ and $x = 2, y = 0$.

SECTION 3.5

1. First, we write the system as

$$x = 4 - 0.5y$$
$$y = 2 + 0.75x$$

It is now straightforward to apply iterative methods. The Jacobi method gives the following results.

Iteration	x	y
0	2	3
1	2.5	3.5
2	2.25	3.875
3	2.0625	3.6875
4	2.15625	3.546875
5	2.2265625	3.6171875
Error	0.04444	-0.01917

The Gauss–Seidel method gives.

Iteration	x	y
0	2	3
1	2.5	3.875
2	2.0625	3.546875
3	2.2265625	3.66992187
4	2.16503906	3.6237793
5	2.18811035	3.64108276
Error	0.006292	0.004719

SECTION 4.1

1. The following table summarizes the calculations necessary for interval estimates of the gradient at different points on the function

X	$f(x)$	$f(x+0.01)$	$\frac{f(x+0.01)\quad f(x)}{0.01}$
1	1	1.03	3.03
2	8	8.121	12.06
3	27	27.271	27.09

This illustrates the property that the gradient changes at different points on a non-linear function.

SECTION 4.2

1. Let $\frac{\Delta y}{\Delta x} = \frac{\sqrt{x+\Delta x} - \sqrt{x}}{\Delta x}$, multiplying numerator and denominator by $\sqrt{x+\Delta x} + \sqrt{x}$ gives

$$\frac{\Delta y}{\Delta x}=\frac{\sqrt{x+\Delta x}-\sqrt{x}}{\Delta x}\times\frac{\sqrt{x+\Delta x}+\sqrt{x}}{\sqrt{x+\Delta x}+\sqrt{x}}=\frac{x+\Delta x-x}{\Delta x\sqrt{x+\Delta x}+\Delta x\sqrt{x}}$$
$$=\frac{1}{\sqrt{x+\Delta x}+\sqrt{x}}$$

Therefore, we have $f'(x)=\dfrac{1}{\mathrm{st}\left(\sqrt{x+\Delta x}+\sqrt{x}\right)}=\dfrac{1}{2\sqrt{x}}$, which is the required result. Note that this is not defined for $x=0$.

SECTION 4.3

1. $y=4x(x+1)^2$ therefore $\dfrac{dy}{dx}=4x.2(x+1)+(x+1)^2\,4$
   $=8x(x+1)+4(x+1)^2$.
3. $y=\sqrt{(4x^2+2x)}$ therefore $\dfrac{dy}{dx}=\dfrac{(4x+1)}{\sqrt{(4x^2+2x)}}$.

SECTION 4.4

1. We have $q=60-3p$. The inverse demand function is therefore given by $p=20-q/3$. We can therefore write the elasticity function as

$$\eta^D=-(-3)\times\frac{20-q/3}{q}=3\left(\frac{20}{q}-\frac{1}{3}\right)$$

It follows that demand is price elastic if

$$3\left(\frac{20}{q}-\frac{1}{3}\right)>1\Leftrightarrow\frac{20}{q}-\frac{1}{3}>\frac{1}{3}\Leftrightarrow 20-\frac{q}{3}>\frac{q}{3}$$
$$\Leftrightarrow 60>2q\Leftrightarrow q<30$$

Therefore, demand is price elastic in the range $0\le q<30$ and price inelastic in the range $30<q\le 60$.

SECTION 4.5

1. (a) $y = -\frac{a}{x^2}$ $\frac{dy}{dx} = \frac{2a}{x^3}$ $\frac{d^2y}{dx^2} = -\frac{6a}{x^4}$

   (b) $y = \exp(2x)$ $\frac{dy}{dx} = 2\exp(2x)$ $\frac{d^2y}{dx^2} = 4\exp(2x)$

   (c) $y = 3\ln(x)$ $\frac{dy}{dx} = \frac{3}{x}$ $\frac{d^2y}{dx^2} = -\frac{3}{x^2}$

SECTION 4.6

1. The forward difference estimate is given by the expression

$$\frac{f(x+h) - f(x)}{h}$$

where h is a small increment. Using a Taylor series expansion around $h = 0$ we have

$$f(x+h) = f(x) + f'(x)h + \frac{1}{2!}f''(x)h^2 + \text{higher order terms.}$$

Substituting this into the expression for the forward difference estimate and rearranging gives

$$\begin{aligned}\frac{f(x+h) - f(x)}{h} &= f'(x) + \frac{1}{2!}f''(x)h + \text{higher order terms}\\ &= f'(x) + O(h)\end{aligned}$$

which is the required result.

SECTION 5.1

1. (a) The first derivative is $f'(x) = 8x + 2$. Therefore, there is an interior critical point at $x = -1/4$ and, since the second derivative is $f''(x) = 8$, this is a local minimum with $f(-1/4) = -1/4$. The end points are $f(-1) = 2$ and $f(2) = 20$. Hence, the interior critical point is also the global minimum with the global maximum occurring at the upper end point.

(b) The first derivative is $f'(x) = 3x^2 - 12$. Therefore, there are interior critical points at $x = 2$ and $x = -2$. The second derivative is $f''(x) = 6x$ and therefore, $x = 2$ is a local minimum with $f(2) = -16$, and $x = -2$ is a local maximum, with $f(-2) = 16$. The end points have value $f(-5) = -65$ and $f(5) = 65$. Therefore, the global minimum and maximum values occur at the end points of the function.

(c) The first derivative is $f'(x) = 2x^2 - 2$. Therefore, there are local stationary points at $x = 1$ and $x = -1$. The second derivative is $4x$ and therefore $x = 1$ is a local minimum with $f(1) = -4/3$ and $x = -1$ is a local maximum with $f(-1) = 4/3$. The end points are $f(-2) = -4/3$ and $f(2) = 4/3$, which are the same values as at the local turning points. Hence the global maximum is 4/3 but this occurs at two different values of x. Similarly, the global minimum is -4/3 and this also occurs at two different values of x.

3. The first derivative of this function is $f'(x) = -1/x^2$. There is no interior point at which this is equal to zero. If we evaluate the end points however, then $f(1) = 1$, which is the global maximum, and 0 provides a lower bound as $x \to \infty$, meaning that this is the infimum of the function.

SECTION 5.2

1. The profit function can be written

$$\Pi(q) = (72 - 2q)q - 10q^2 = 72q - 12q^2.$$

The first-order condition for a maximum is therefore

$$\frac{d\Pi}{dq} = 72 - 24q = 0 \Rightarrow q = 3.$$

We can show that this is a maximum because $\frac{d^2\Pi}{dq^2} = -24 < 0$.

SECTION 5.3

1. $y = x^2$ is strictly convex if and only if

$$\lambda x_1^2 + (1-\lambda)x_2^2 > (\lambda x_1 + (1-\lambda)x_2)^2$$
$$\lambda x_1^2 + x_2^2 - \lambda x_2^2 > \lambda^2 x_1^2 + x_2^2 - 2\lambda x_2^2 + \lambda^2 x_2^2 + 2\lambda x_1 x_2 - 2\lambda^2 x_1 x_2$$

Subtract x_2^2 and add $2\lambda x_2^2$ from both sides

$$\lambda x_1^2 + \lambda x_2^2 > \lambda^2 x_1^2 + \lambda^2 x_2^2 + 2\lambda x_1 x_2 - 2\lambda^2 x_1 x_2$$
$$\left(\lambda - \lambda^2\right) x_1^2 + \left(\lambda - \lambda^2\right) x_2^2 > 2\lambda x_1 x_2 - 2\lambda^2 x_1 x_2$$
$$\lambda\left(1-\lambda\right)\left(x_1^2 + x_2^2\right) > 2\lambda\left(1-\lambda\right) x_1 x_2$$
$$x_1^2 + x_2^2 - 2x_1 x_2 > 0$$
$$\left(x_1 - x_2\right)^2 > 0$$

which is obviously true.

Note that this is much more easily demonstrated using the second-order derivative condition since $f''(x) = 2 > 0$.

SECTION 5.4

1. For the function $f(x) = x^3/3 - 4x + 1$ we have $f'(x) = x^2 - 4$. Therefore, $f'(x_L) = f'(1) = -3$ and $f'(x_U) = f'(5) = 21$. Since the first derivative changes signs, it follows that there is a stationary point between these two points and, since it changes from negative to positive, it follows that there is a minimum point.

   If we set $x_M = (1+5)/2 = 3$ $x_M = (1+5)/2 = 3$, then $f'(3) = 5$. Therefore, we set $x_U = 3$ and recalculate $x_M = (1+3)/2 = 2$. We now have $f'(x_M) = f'(2) = 0$. Therefore, we have located the stationary point on the second iteration.

SECTION 6.1

1. For the function $z = f(x,y) = ax^3 + by^2$, we have $f(\lambda x, \lambda y) = a\lambda^3 x + b\lambda^2 y$ it is not possible to factorize this such that $f(\lambda x, \lambda y) = \lambda^r f(x,y)$ and therefore this function is not homogeneous. This illustrates an important point. In general, when functions have different powers of x and y on the right hand side, then they will not be homogeneous functions.

3. Consider the general Cobb-Douglas production function $Y = AK^a N^\beta$. For this to exhibit constant returns to scale, the exponents must sum to one. That is, we require $\beta = 1 - a$, which means that we can write the function as

$$Y = AK^a N^{1-a}.$$

Dividing both sides by N gives

$$\frac{Y}{N} = AK^{a} N^{-a} = A\left(\frac{K}{N}\right)^{a}.$$

That is, output per capita is a function of capital input per capita.

SECTION 6.2

1. **(a)** $f_x = 3x^2 / y \quad f_y = -x^3 / y^2$

   (b) $f_x = \exp(y) \quad f_y = x\exp(y)$

   (c) $f_x = 6x\left(x^2 + y^2\right)^2 \quad f_y = 6y\left(x^2 + y^2\right)^2$

3. For the function $Y = F(K,N) = K^{a} N^{1-a}$, the first-order partial derivatives are

$$F_K = a\, K^{a-1} N^{1-a} \quad F_N = (1-a)K^{a} N^{-a}$$

These are both positive because both K and N can only take on positive values, and we have assumed that $0 < a < 1$.

The second-order partial derivatives are given by

$$F_{KK} = (a-1)a\, K^{a-2} N^{1-a} \quad F_{NN} = -a\,(1-a)K^{a} N^{-a-1}$$

In this case, the assumption that $0 < a < 1$ means that both of these second-order partial derivatives are negative. This function is, therefore, consistent with the assumptions that the marginal products of capital and labor are positive but diminishing as more of one factor is added to a fixed quantity of the other.

SECTION 6.3

1. **(a)** $dz = (6x + 4y)dx + \left(6y^2 + 4x\right)dy$

   (b) $dz = \ln y\, dx + \frac{x}{y} dy$

   (c) $dz = \exp(x-y)dx - \exp(x-y)dy$

3. We have $u(c_1,c_2)=\ln(c_1)+\beta\ln(c_2)$ and therefore.

$$du=\frac{1}{c_1}dc_1+\frac{\beta}{c_2}dc_2.$$

Setting $du=0$ allows us to solve for the gradient of the indifference curves as

$$\frac{dc_2}{dc_1}=-\left(\frac{1}{\beta}\right)\frac{c_2}{c_1}$$

These curves have the standard shape as those shown in the text in that they approach the horizontal axis asymptotically as $c_1\to\infty$ and the vertical axis asymptotically as $c_1\to 0$.

SECTION 6.4

1. We have $z(x,y)=x^2+2y^2+2x-4xy$. The first-order conditions for a turning point are

$$\frac{\partial z}{\partial x}=2x+2-4y=0$$
$$\frac{\partial z}{\partial y}=4y-4x=0$$

From the second equation we have $y=x$. Substituting this into the first equation gives $2-2x=0$. Therefore, the solution is $y=x=1$.

The second-order derivatives and the cross partial derivative are

$$\frac{\partial^2 z}{\partial x^2}=2\ \frac{\partial^2 z}{\partial y^2}=4\ \frac{\partial^2 z}{\partial x\partial y}=-4$$

We have

$$\frac{\partial^2 z}{\partial x^2}\frac{\partial^2 z}{\partial y^2}-\left(\frac{\partial^2 z}{\partial x\partial y}\right)^2=2\times 4-(-4)^2=-8$$

It therefore follows that this is a saddle-point.

SECTION 6.5

1. The contours of the function are defined by the equation $\bar{z} = 2x^2 + y^2$, where $\bar{z}$ is a constant. Using the implicit function rule, we have

$$4x + 2y\frac{dy}{dx} = 0 \Rightarrow \frac{dy}{dx} = -\frac{2x}{y}$$

Note that this is always negative from the assumption that both x and y are positive real numbers. Differentiating again gives us

$$\frac{d^2y}{dx^2} = -2\left(\frac{y - x\,dy/dx}{y^2}\right)$$

Since dy/dx is always negative, it follows that the term in parentheses is always positive, and therefore $d^2y/dx^2 < 0$. This is a sufficient condition to prove the statement that the contours are strictly concave.

3. To minimize the costs of production, we set up the following Lagrangian function

$$L(N, K, \lambda) = 2N + 0.5K - \lambda\left(N^{0.5}K^{0.5} - 100\right)$$

which gives us the following first-order conditions.

$$\frac{\partial L}{\partial N} = 2 - \lambda\frac{N^{-0.5}K^{0.5}}{2} = 0$$
$$\frac{\partial L}{\partial K} = 0.5 - \lambda\frac{N^{0.5}K^{-0.5}}{2} = 0$$
$$\frac{\partial L}{\partial \lambda} = N^{0.5}K^{0.5} - 100 = 0$$

From the first two equations, we have

$$\lambda = \frac{4}{N^{-0.5}K^{0.5}} = \frac{1}{N^{0.5}K^{-0.5}} \Rightarrow 4N = K$$

Substituting into the third equation gives $N^{0.5}(4N)^{0.5} = 100$ or $2N = 100 \Rightarrow N = 50$ which, in turn, gives us $K = 4N = 200$.

SECTION 6.6

1. The Hessian matrix is defined as $H = \begin{bmatrix} \partial^2 z / \partial x^2 & \partial^2 z / \partial x \partial y \\ \partial^2 z / \partial x \partial y & \partial^2 z / \partial y^2 \end{bmatrix}$. The conditions for a local maximum are

$$(1) \qquad \frac{\partial^2 z}{\partial x^2}\frac{\partial^2 z}{\partial y^2} - \left(\frac{\partial^2 z}{\partial x \partial y}\right)^2 > 0$$

$$(2) \qquad \frac{\partial^2 z}{\partial x^2} < 0$$

The first condition simply states that the determinant must be positive. Note that for this to hold, then both second-order partial derivatives must have the same sign. It therefore follows that, if the second condition holds, then the trace of the Hessian matrix must be negative. Therefore, the conditions $tr(H)$ and $\det(H) > 0$ are exactly equivalent to the standard second-order conditions for a local maximum.

SECTION 7.1

1. **(a)** $\sum_0^1 3x^2 \Delta x = \frac{1}{4}\left(0 + \frac{3}{16} + \frac{3}{4} + \frac{27}{16}\right) = 0.65625$

   (b) $\sum_{-1}^0 3x^2 \Delta x = \frac{1}{4}\left(3 + \frac{27}{16} + \frac{3}{4} + \frac{3}{16}\right) = 1.40625$

   (c) $\sum_0^1 (x-1)\Delta x = -\frac{1}{4}\left(1 + \frac{3}{4} + \frac{1}{2} + \frac{1}{4}\right) = -0.625$

The answer to part 1(c) is negative because this curve lies below the x-axis in the interval $[0,1]$. This is interpreted as a negative area when we calculate the definite integral. In parts (a) and (b), we always have $f(x) \geq 0$, and so this complication does not arise.

SECTION 7.2

1. We can differentiate the function $F(x) = x\ln x - x + C$ using the product rule to obtain

$$F'(x) = x \times \frac{1}{x} + \ln x - 1 = \ln x$$

which is the required result.

3. **(a)** $\int_0^2 (x^2+4x)\,dx = \left[\frac{x^3}{3}+2x^2\right]_0^2 = \left(\frac{8}{3}+8\right)-(0) = \frac{32}{3}$

(b) $\int_{-1}^1 x^3 dx = \left[\frac{x^4}{4}\right]_{-1}^1 = \left(\frac{1}{4}\right)-\left(\frac{1}{4}\right)=0$

(c) $\int_{-\infty}^0 \exp(x)\,dx = [\exp(x)]_{-\infty}^0 = 1-\lim_{x\to-\infty}\exp(x)=1$

SECTION 7.3

1. We wish to calculate $\int \exp(ax)\,dx$. Let $u=ax$, and, since $du=adx$, we can write the integral as $\int \exp(u)/a\,du=(1/a)\exp(u)+C=(1/a)$ $\exp(ax)+C$, which is the required result.

3. We wish to calculate $\int x\exp(x)\,dx$. Let $u=x$ and $v=\exp(x)$. Using the method of integration by parts, we have

$$\int x\exp(x)\,dx = x\exp(x)-\int \exp(x)\,dx=\exp(x)(x-1)+C$$

SECTION 7.4

1. The income stream is given by the following equation $100\exp(-0.15t)$. It is discounted at rate 5% per annum and therefore the present value of the income stream is given by the following integral.

$$\int_0^\infty 100\exp(-0.15t)\exp(-0.05t)\,dt = \int_0^\infty 100\exp(-0.2t)\,dt$$

$$=\left[-\frac{1}{0.2}\times 100\exp(-0.2t)\right]_0^\infty = -\frac{100}{0.2}[\exp(-0.2t)]_0^\infty$$

$$=-(0)-\left(\frac{100}{0.2}\right)=500$$

Therefore, the present value of the income stream described is $2,000.

3. If the inverse demand curve is $p = 4 - 2q$, and the equilibrium price is 2, then the equilibrium quantity is $q = 1$. The total area under the demand curve up to this point is

$$\int_0^1 (4-2q)\,dq = \left[4q - q^2\right]_0^1 = 3$$

Consumer surplus is equal to this area minus the amount consumers pay for the product. Since the market price is 2 and the quantity is 1, we have consumer surplus equal to $3 - 2 \times 1 = 1$.

SECTION 7.5

(a) We have $\int_0^1 (5x+2)\,dx = \left[5x^2/2 + 2x\right]_0^1 = 9/2$, and using Simpson's rule, we have the following approximation

$$\frac{1}{6}\left(f(0) + 4f\left(\frac{1}{2}\right) + f(1)\right) = \frac{1}{6}\left(2 + 4\times\left(\frac{5}{2}+2\right) + 7\right) = \frac{27}{6} = \frac{9}{2}$$

(b) We have $\int_0^1 \left(2x^2 + 3x\right)dx = \left[\frac{2x^3}{3} + \frac{3x}{2}\right]_0^1 = \frac{2}{3} + \frac{3}{2} = \frac{13}{6}$.

Using Simpson's rule we have

$$\frac{1}{6}\left(0 + 4\times\left(\frac{1}{2}+\frac{3}{2}\right) + 5\right) = \frac{13}{6}$$

(c) We have $\int_0^1 2x^3 = \left[\frac{x^4}{2}\right]_0^1 = \frac{1}{2}$

Using Simpson' rule, we have

$$\frac{1}{6}\left(0 + 4\times\frac{2}{8} + 2\right) = \frac{1}{2}$$

(d) We have $\int_0^1 5x^4 dx = \left[x^5\right]_0^1 = 1$.

Using Simpson's rule we have

$$\frac{1}{6}\left(0 + 4\times\frac{5}{2^4} + 5\right) = \frac{25}{24} = 1.0416$$

This demonstrates that Simpson's rule is exact for polynomials up to, and including, order 3 but no longer holds for polynomials of order 4 and higher.

SECTION 8.1

1. **(a)** $AB = \begin{bmatrix} 16 & 30 \\ 6 & 12 \end{bmatrix}$

(b) $AB = \begin{bmatrix} 6 & 3 \\ 8 & 4 \end{bmatrix}$

(c) $AB = 19$

SECTION 8.2

1. Let $A = \begin{bmatrix} a_{11} & ka_{11} \\ a_{12} & ka_{12} \end{bmatrix}$ where k is any real number. The determinant of this matrix is

$$\det(A) = a_{11}ka_{12} - ka_{11}a_{12} = k(a_{11}a_{12} - a_{11}a_{12}) = 0$$

for any value of k. This establishes the result.

SECTION 8.3

1. To demonstrate that this statement is true, we will show that the product of the original matrix and its proposed inverse is equal to the identity matrix. We have

$$\begin{bmatrix} a_{11} & a_{12} \\ a_{21} & a_{22} \end{bmatrix} \cdot \frac{1}{\Delta}\begin{bmatrix} a_{22} & -a_{12} \\ -a_{21} & a_{11} \end{bmatrix} = \frac{1}{\Delta}\begin{bmatrix} a_{11}a_{22} - a_{21}a_{12} & -a_{11}a_{12} + a_{11}a_{12} \\ a_{22}a_{21} - a_{21}a_{22} & -a_{12}a_{21} + a_{11}a_{22} \end{bmatrix}$$

It is immediately obvious that the off-diagonal elements are equal to zero and that both of the diagonal elements can be written as

$$\frac{a_{11}a_{22} - a_{12}a_{22}}{\Delta}.$$

Since $\Delta = a_{11}a_{22} - a_{12}a_{21}$, it follows that both diagonal elements are equal to one. Therefore, the product of these two matrices is the identity matrix and that the second matrix is the inverse of the first matrix.

SECTION 8.4

1. The computer code gives us the following inverse for the matrix in the question

```
The matrix inverse is

  0.1856      0.1546      -0.1237      0.0103
  0.1340      0.2784       1.5773     -1.3814
  0.1649     -0.1959      -0.4433      0.4536
 -0.1753      0.0206      -0.2165      0.2680
```

SECTION 8.5

1. To solve for the eigenvalues, we find the roots of the characteristic equation defined by

$$\begin{vmatrix} 3-\lambda & 1 \\ 0 & 2-\lambda \end{vmatrix} = 0$$

which, in this case, are obviously $\lambda_1 = 3$ and $\lambda_2 = 2$.

To solve for the eigenvector associated with $\lambda_1 = 3$, we look for v_1 and v_2 such that

$$\begin{bmatrix} 3 & 1 \\ 0 & 2 \end{bmatrix}\begin{bmatrix} v_1 \\ v_2 \end{bmatrix} = \begin{bmatrix} 3v_1 \\ 3v_2 \end{bmatrix}$$

We have $3v_1 + v_2 = 3v_1$ which implies $v_2 = 0$. Normalizing so that the modulus is equal to one, means that the eigenvector associated with $\lambda_1 = 3$ is $\begin{bmatrix} 1 & 0 \end{bmatrix}^T$.

To solve for the eigenvector associated with $\lambda_2 = 2$, we look for v_1 and v_2 such that

$$\begin{bmatrix} 3 & 1 \\ 0 & 2 \end{bmatrix}\begin{bmatrix} v_1 \\ v_2 \end{bmatrix} = \begin{bmatrix} 2v_1 \\ 2v_2 \end{bmatrix}$$

We have $3v_1 + v_2 = 2v_1$ which implies $v_2 = -v_1$. This means that the eigenvector can be written as $\begin{bmatrix} 1 & -1 \end{bmatrix}^T$. Alternatively, normalizing so that the modulus is equal to one, means that the eigenvector associated with $\lambda_2 = 2$ is $\left[1/\sqrt{2} \quad -1/\sqrt{2}\right]^T$.

SECTION 9.1

1. **(a)** We have $\dfrac{dy}{dx} = \dfrac{xy^2}{(1+x)}$, which means we can write

$$\int \frac{1}{y^2} dy = \int \frac{x}{(1+x)} dx$$
$$-y^{-1} = x - \ln(1+x) + C$$
$$y = \frac{1}{\ln(1+x) - x - C}$$

Using the initial condition, we have $1 = 1/(-C) \Rightarrow C = -1$ and the solution is $y(x) = 1/\left(1 - x + \ln(1+x)\right)$.

(b) We have $\dfrac{dy}{dx} = e^{-y}(3x-1)$, which means we can write

$$\int e^y dy = \int (3x-1)dx$$
$$e^y = \frac{3x^2}{2} - x + C$$
$$y = \ln\left(\frac{3x^2}{2} - x + C\right)$$

Using the initial condition, we have $0 = \ln(C) \Rightarrow C = 1$ and the solution is $y(x) = \ln\left(3x^2/2 - x + 1\right)$.

3. This differential equation is separable, therefore it can be written as

$$\exp(y)dy = \left(2x - \frac{3}{2}\right)dx.$$

Integrating gives us the solution

$$\exp(y) = x^2 - \frac{3}{2}x + C$$

where C is the constant of integration and, from the initial condition, we have $\exp(0) = C = 1$. Therefore, the particular solution takes the form

$$\exp(y) = x^2 - \frac{3}{2}x + 1$$
$$y(x) = \ln\left(x^2 - \frac{3}{2}x + 1\right)$$

This solution is only valid if $g(x) = x^2 - (3/2)x + 1 > 0$. We know that $g(0) > 0$ and we can show that this function has complex roots since $(3/2)^2 - 4 < 0$. It follows that $g(x)$ has no zeroes in the class of real numbers and it is therefore always positive for $x > 0$.

SECTION 9.2

1. To solve this equation by separation of variables, we first define $u = y - 4$. Substituting this into the equation given allows us to write as a homogeneous equation of the form

$$\frac{du}{dx} - u = 0$$

This can be solved easily by separation of variables

$$\int \frac{1}{u} du = \int 1\, dx \Rightarrow \ln(u) = x + C_1$$

Substituting back for u and exponentiating this equation gives us the general solution

$$\ln(y-4) = x + C_1 \Rightarrow y(x) = C_2 \exp(x) + 4 \text{ where } C_2 = \exp(C_1)$$

An alternative method of solution is to add the complementary function and the particular integral. For this problem, we have

$$y_c(x) = C\exp(x)$$
$$y_p(x) = 4$$
$$y_g(x) = C\exp(x) + 4$$

Therefore, we get the same answer whichever method we use.

3. **(a)** $y_g(x) = C\exp(2x) + 2$

$y(0) = 1 \Rightarrow 1 = C + 2 \Rightarrow C = -1$

$y(x) = 2 - \exp(2x)$

(b) $y_g(x) = C\exp(-3x) + 1$

$y(-1) = 2 \Rightarrow 2 = C\exp(3) + 1 \Rightarrow C = 2\exp(-3)$

$y(x) = 1 + 2\exp(-3x - 3)$

(c) $y_g(x) = C\exp(-0.1x) + 20$

$y(1) = 5 \Rightarrow 5 = C\exp(-0.1) + 20 \Rightarrow C = -15\exp(0.1)$

$y(x) = 20 - 15\exp(-0.1x + 0.1)$

SECTION 9.3

1. **(a)** The integrating factor is $v(x) = \exp\left(\int 0.5dx\right) = \exp(0.5x)$. Therefore, we have

$$\frac{d(y\exp(0.5x))}{dx} = 0$$
$$y\exp(0.5x) = C$$
$$y_g(x) = C\exp(-0.5x)$$

(b) The integrating factor is $v(x) = \exp\left(\int\left(\frac{4}{x}\right)dx\right) = \exp(4\ln x) = x^4$. Therefore, we have

$$\frac{d(yx^4)}{dx} = 0$$
$$yx^4 = C$$
$$y_g(x) = Cx^{-4}$$

(c) The equation can be written as $dy/dx + (5/x^2)y = 0$ and therefore the integrating factor is $v(x) = \exp\left(\int\frac{5}{x^2}dx\right) = \exp\left(-\frac{5}{x}\right)$. Therefore, we have

$$\frac{d\left(y\exp(-5/x)\right)}{dx} = 0$$
$$y\exp(-5/x) = C$$
$$y_g(x) = C\exp(5/x)$$

3. The constant of integration for this problem is $\exp(ax)$. We therefore have

$$\frac{d\left(y\exp(ax)\right)}{dx} = b\exp(ax)$$

Integrating yields

$$y\exp(ax) = \frac{b}{a}\exp(ax) + C$$

which can be solved to give the following equation for the general solution

$$y_g(x) = \frac{b}{a} + C\exp(-ax)$$

This is equal to the sum of the general solution for the associated homogeneous equation and the particular integral. It therefore confirms that the method demonstrated in Section 9.2 is correct.

SECTION 9.4

1. The complementary function is given by the solution of the homogeneous equation obtained by setting the right-hand side equal to zero. We have

$$y_c(x) = C\exp(-2x)$$

To find a particular integral, we guess a solution of the form $y_p(x) = a + bx$. This gives us

$$b + (2a + bx) = 3x$$
$$(b + 2a) + 2bx = 3$$

We therefore have $b = 3/2$ and $a = -3/4$. The general solution for the non-homogeneous equation is therefore.

$$y_g(x) = C\exp(-2x) - \frac{3}{4} + \frac{3}{2}x$$

where C is a constant of integration.

3. The complementary function is given by the solution of the homogeneous equation obtained by setting the right-hand side equal to zero. We have

$$y_c(x) = C\exp(-0.5x)$$

To find a particular integral, we guess a solution of the form $y_p(x) = A\exp(bx)$. This gives us

$$bA\exp(bx) + A\exp(bx) = \exp(0.5x)$$
$$(bA + A)\exp(bx) = \exp(0.5x)$$

We therefore have $b = 0.5$ and $A = 1$. The general solution for the non-homogeneous equation is therefore.

$$y_g(x) = C\exp(-0.5x) + \exp(0.5x)$$

where C is a constant of integration. To solve for the constant of integration, we use the initial condition $10 = C + 1$. We therefore have $C = 9$ and the particular solution of the differential equation with the initial condition we are given is

$$y(x) = 9\exp(-0.5x) + \exp(0.5x).$$

SECTION 9.5

1. First, we note that the differential equation $dy/dx = -0.2y$ with $y(0) = 1$ can be solved analytically to give $y(x) = \exp(-0.2x)$. Therefore, we can obtain an exact solution for $y(1) = 0.81873$ (to five decimal places).

Now applying Euler's method, we have

$$x_{i+1} = x_i + h$$
$$y_{i+1} = y_i - 0.2y_i h = y_i(1 - 0.2h)$$

If $h = 0.5$, then we have

x	Y
0	1
0.5	0.9
1.0	0.81

If $h = 0.2$, then we have

x	y
0	1
0.2	0.96
0.4	0.9216
0.6	0.8847
0.8	0.8493
1.0	0.8153

Note that, by reducing the interval h, we increase the accuracy of the calculation.

SECTION 9.6

1. In the Cagan model, the solution for the price level is given by the equation $P(t) = M(t)\exp(a\sigma)$, where σ is the rate of growth of the money stock. Let M_0 be the value of the money stock at t_0. If the initial growth rate is equal to σ_1 and there is an instantaneous cut in this to $\sigma_2 < \sigma_1$, then the price level falls from $M_0 \exp(a\sigma_1)$ to $M_0 \exp(a\sigma_2)$. It will then continue to grow at the lower rate σ_2.

SECTION 10.1

1. The characteristic equation takes the form $\lambda^2 + 2\lambda + 5/4 = 0$ which gives us roots

$$\lambda_{1,2} = \frac{-2 \pm \sqrt{4-5}}{2} = -1 \pm i$$

The general solution of the differential equation, therefore, takes the form

$$y_g(x) = \exp(-x)\{C_1 \cos(x) + C_2 \sin(x)\}$$

3. The characteristic equation takes the form $\lambda^2 - 10\lambda + 25 = 0$ which gives us roots

$$\lambda_{1,2} = \frac{10 \pm \sqrt{100-100}}{2} = 5$$

The general solution of the differential equation therefore takes the form

$$y_g(x) = C_1 \exp(5x) + C_2 x \exp(5x)$$

SECTION 10.2

1. Dividing through by 8 gives the following characteristic equation

$$\lambda^2 + \frac{3}{4}\lambda + \frac{1}{8} = 0$$
$$\lambda = \frac{-3/4 \pm \sqrt{9/16 - 1/2}}{2} = -\frac{1}{4} \text{ or } -\frac{1}{2}$$

We can therefore write the general solution as

$$y_g(x) = C_1 \exp\left(-\frac{1}{2}x\right) + C_2\left(-\frac{1}{4}x\right)$$

To apply the initial conditions, we differentiate this expression to get

$$\frac{dy_g}{dx} = -\frac{1}{2}C_1 \exp\left(-\frac{1}{2}x\right) - \frac{1}{4}C_2 \exp\left(-\frac{1}{4}x\right)$$

We can now apply the initial conditions to get

$$C_1 + C_2 = 2$$
$$-\frac{1}{2}C_1 - \frac{1}{4}C_2 = 0$$

which solve to give us $C_1 = -2$ and $C_2 = 4$. The particular solution consistent with these initial conditions is therefore

$$y(x) = -2\exp\left(-\frac{1}{2}x\right) + 4\exp\left(-\frac{1}{4}x\right).$$

3. Dividing through by 9 gives us the following characteristic equation

$$\lambda^2 + \frac{2}{3}\lambda + \frac{1}{9} = 0$$
$$\lambda = \frac{-2/3 \pm \sqrt{4/9 - 4/9}}{2} = -\frac{1}{3}$$

Since we have a repeated root, the general solution takes the form

$$y_g(x) = C_1 \exp\left(-\frac{1}{3}x\right) + C_2 x \exp\left(-\frac{1}{3}x\right)$$

The initial conditions give us

$$C_1 + C_2 = 3$$
$$\exp(-3)\{C_1 - C_2\} = 0$$

which solve to give us $C_1 = C_2 = 3/2$. Therefore the particular solution which is consistent with these initial conditions is given by

$$y(x) = \frac{3}{2}\exp\left(-\frac{1}{3}x\right) + \frac{3}{2}x\exp\left(-\frac{1}{3}x\right).$$

SECTION 10.3

1. The complementary function is the solution of the homogeneous equation and is the same for all three cases. The characteristic equation is $\lambda^2 + 3\lambda + 2 = 0$ which factorizes to give $(\lambda + 2)(\lambda + 1) = 0$ and therefore the roots are $\lambda_1 = -2$ and $\lambda_2 = -1$. The complementary function is therefore given by

$$y_c(x) = C_1 \exp(-2x) + C_2 \exp(-x)$$

In each case we now need to calculate a particular integral $y_p(x)$. We can then calculate the general solution to the non-homogeneous equation as $y_g(x) = y_c(x) + y_p(x)$.

(a) We assume a solution of the form $y_p(x) = a + bx$. This gives $3b + 2(a + bx) = 2 + 3x$. Equating coefficients then gives $b = 3/2$ and $a = -5/4$. The particular solution is therefore given by $y_p(x) = -(5/4) + (3/2)x$.

(b) We assume a solution of the form $a + bx + cx^2$. This gives $2c + 3(b + 2cx) + 2(a + bx + cx^2) = 4x^2$. Equating coefficients gives $c = 2$, $b = -6$ and $a = 7$. The particular solution is therefore given by $y_p(x) = 7 - 6x + 2x^2$.

(c) We assume a solution of the form $A\exp(bx)$. This gives $\exp(bx)\{b^2 A + 3bA + 2A\} = 2\exp(x/2)$. Equating coefficients gives $b = 1/2$ and $A = 8/15$. The particular solution is therefore given by $y_p(x) = (8/15)\exp(x/2)$.

SECTION 10.4

1. **(a)** We have $d^2y/dx^2 + 3x\,dy/dx + 2y = x$. Define $z = dy/dx$, this allows us to write the equation as

$$\frac{dz}{dx} = -3xz - 2y + x$$
$$\frac{dy}{dx} = z$$

(b) We have $4xd^2y/dx^2 - 2y = \exp(x)$. Define $z = dy/dx$, this allows us to write the equation as

$$\frac{dz}{dx} = \frac{1}{2x}y + \frac{\exp(x)}{4x}$$
$$\frac{dy}{dx} = z$$

SECTION 11.1

1. **(a)** The general solution of the homogeneous equation is $\tilde{y}_n = C(2)^n$ and the particular integral is $y_p = -4$. Therefore, the general solution of the non-homogeneous equation is $y_n = C(2)^n - 4$.

(b) The general solution of the homogeneous equation is $\tilde{y}_n = C\left(-\frac{1}{2}\right)^n$ and the particular integral is $y_p = 4/3$. Therefore, the general solution of the non-homogeneous equation is $y_n = C\left(-\frac{1}{2}\right)^n + \frac{4}{3}$.

(c) The general solution of the homogeneous equation is $\tilde{y}_n = C(-3)^n$ and the particular integral is $y_p = \frac{1}{4}$. Therefore, the general solution of the non-homogeneous equation is $y_n = C(-3)^n + \frac{1}{4}$.

3. The general solution for the homogeneous equation is $\tilde{y}_n = C\left(\frac{1}{4}\right)^n$. To solve for the particular integral, we note that the non-homogeneous part of the equation is an exponential function. We, therefore, assume a function of the form $y_p = A\exp(bn)$, where A and b are unknown parameters. Using the method of undetermined coefficients, we have

$$A\exp(bn) - \frac{1}{4}A\exp(bn)\exp(-b) = \exp\left(-\frac{1}{2}n\right)$$

It follows immediately that $b = -1/2$. We therefore have

$$A - \frac{1}{4}A\exp\left(\frac{1}{2}\right) = 1$$

which can be solved to give us $A = 4/\left(4 - \exp\left(\frac{1}{2}\right)\right)$. The general solution for the non-homogeneous equation is therefore

$$y_n = C\left(\frac{1}{4}\right)^n + \frac{4}{4 - \exp(1/2)}.$$

SECTION 11.2

1. (a) The characteristic equation is $\lambda^2 - \lambda - 2 = 0$ which gives us roots $\lambda_1 = -1$ and $\lambda_2 = 2$. Therefore the general solution takes the form

$$y_n = C_1(-1)^n + C_2(2)^n$$

(b) The characteristic equation is $\lambda^2 + 2\lambda + 5 = 0$ which gives us roots $\lambda_1 = -1 + 2i$ and $\lambda_2 = -1 - 2i$. The modulus is $r = \sqrt{1+4} = \sqrt{5}$ and the argument is $\theta = \tan^{-1}\left(\frac{2}{-1}\right) = -1.107$. Therefore, the general solution takes the form

$$y_n = \left(\sqrt{5}\right)^n \left(C_1 \cos(-1.107n) + C_2 \sin(-1.107n)\right)$$

(c) The characteristic equation is $\lambda^2 - \frac{2}{3}\lambda + \frac{1}{9} = 0$ which gives us repeated roots $\lambda_1 = \lambda_2 = 1/3$. Therefore, the general solution takes the form

$$y_n = (C_1 + C_2 n)\left(\frac{1}{3}\right)^n$$

3. The characteristic equation for this problem is $\lambda^2 - (2/5)\lambda + (1/25) = 0$ which has repeated roots $\lambda_1 = \lambda_2 = 0.2$. The particular integral is $y_p = 125/16$. Therefore, the general solution of the non-homogeneous equation is

$$y_n = (C_1 + C_2 n)(0.2)^n + \frac{125}{16}.$$

From the initial conditions, we have

$$C_1 + \frac{125}{16} = 0$$
$$(C_1 + C_2)\frac{1}{5} + \frac{125}{16} = 1$$

These solve to give us $C_1 = -125/16$ and $C_2 = -105/4$. Therefore, the particular solution which is consistent with these initial conditions is

$$y_n = \left(-\frac{125}{16} - \frac{105}{4}n\right)(0.2)^n + \frac{125}{16}.$$

SECTION 11.3

1. Using backward substitution, we have

$$y_t = 0.2y_{t-1} + 0.8 = 0.8\left(1 + 0.2 + 0.2^2 + \ldots + 0.2^t\right) + 0.2^t y_0$$
$$= 0.8\frac{1 - 0.2^t}{1 - 0.2} + 0.2^t y_0$$

As $t \to \infty$ the first term tends to $0.8/(1-0.2) = 1$ while the second term tends to zero.

SECTION 11.4

1. If the dividend is equal to \$10 and the market rate of return is equal to 0.05, then the market fundamental equity price is $\$10/0.05 = \$200$. If the dividend rises to \$15 then the market fundamental price rises to $\$15/0.05 = \$300$. The equation

$$p_t = C(1.05)^t + \frac{d_1}{r} = C(1.05)^t + 200$$

describes the adjustment of the equity price between the time at which the dividend increase is first anticipated, in this case $t = 0$, and the time it occurs.

(a) At $t = 1$ we need $300 = C(1.05) + 200 \Rightarrow C = 95.24$. The equation for p_t therefore takes the form $p_t = \$95.24(1.05)^t + \$200$ for $0 < t \le 1$. Therefore, the price jumps by \$95.24 at date 0.

(b) At $t = 2$ we need $300 = C(1.05)^2 + 200 \Rightarrow C = 90.70$. The equation for p_t therefore takes the form $p_t = \$90.7(1.05)^t + \$200$ for $0 < t \le 2$. Therefore, the price jumps by \$90.70 at date 0.

(c) At $t = 10$ we need $300 = C(1.05)^{10} + 200 \Rightarrow C = 61.39$. The equation for p_t therefore takes the form $p_t = \$61.39(1.05)^t + \$200$ for $0 < t \le 10$. Therefore, the price jumps by \$61.39 at date 0.

This example illustrates the property that the further in the future the change in the dividend rate is expected to take place, the smaller will be the immediate jump in the equity price.

INDEX